COLOR & LIGHT: INFLUENCES AND IMPACT

By

BARBARA COLBY, ASID

"Something old,

Something new,

Something borrowed,

Something ***blue****."*

"He had a ***yellow*** *streak down his back."*

"Tickled ***pink****."*

*"****Green*** *with envy."*

"I feel ***blue****."*

These descriptive terms are a measure of our understanding that color affects us both psychologically as well as physiologically.

Business, education and medicine have acted upon this understanding by spending millions of dollars on the use of light and color. Industry pays for studies to determine the best color schemes to promote efficiency in offices and factories. Some schools have participated in studies showing the effects of color. Hospitals use light and color to reduce patient anxiety and speed their recovery. The federal government has specific standards for light and color in its buildings, as do local governments.

From prehistoric times until today, our perspective of light and color has changed —— and I'm sure that by the year 2090 we will look back on 1990, as we now look back on the primitive use of light in 1890. Major advances in light sources have been factors in successful integration of color and light in changing environments to accommodate diverse moods and preferences. The principles of light and color form a relationship between correct lighting and our environment, which in turn illuminates all phases of life. But, is this true success —— Or should success for mankind be measured in a different way?

Retail Stores

Color works to differentiate the product in the mind of the consumer. In a retail store, color impact must be made within ten seconds, since that's the amount of time consumers give a product as they browse by. Color can be used to create selling zones, as a dramatic effect on forms or structures, or as an illusion to project an idea. For the retailer, color is the great communicator.

Health Care

In recent years, health care colors have dramatically changed from hospital green, white, yellow and beige, to more stimulating, pleasing colors. In the future hospitals will become acute care centers with a shift in the length of stay declining, so using color as a workable tool to make people feel better more quickly makes sense.

The color message for the patient now and in the future will be to "Get-up and get well quickly". The color "Cardiac Blue" is being used in cardiac units to calm patients. Colors, such as aquamarine, a peachy orange, and dogwood pink help to cheer the patient on to good health. The wrong color reflecting on the skin's surface will change the look of skin tones. The new colors are not only get-well colors, but tend to accent the skin's own natural pigmentation. When people look better, they feel better. Because color has the ability to stimulate the autonomic nervous system as well as the brain and body's muscle activity, it is a logical tool to use in health care.

Bridges

Many tall iron bridges are noted for suicides. In London, the Blackfriars bridge, made of black ironwork was repainted a green color. Suicides declined more than thirty percent. Perhaps San Francisco should take note, since the Golden Gate Bridge is notorious for a multitude of suicides and is painted red.

CHAPTER 4

FAIL SAFE COLOR SELECTION

In the early eighties, women began appearing in retail establishments, holding swatches of colored material, while looking for "just the right" color in make-up and clothing for their spring, summer, winter or fall personal palette. At social gatherings, instead of the talk turning to, "What's your astrological sign?", suddenly conversations enveloped the topic of "What color season are you?" Obviously there has been a color revolution going on. Gone are the days of any type of color selection without guidelines.

In spite of our color renaissance, the average person feels color information is very specialized and beyond comprehension. People combine colors expecting them to harmonize, and the color combinations fail. Wasn't that neutral beige supposed to work with brown? Why didn't the grey we selected look right with that red? Isn't grey, or for that matter beige, a neutral? Shouldn't white work with everything? Obviously, there are no universal neutrals in dealing with color harmony. Once you work with color harmonies on a consistent basis, you become aware of this fact very quickly.

If this information is correct, then how does one arrive at a formula for making correct color decisions for harmonious color combinations? In spite of the color information presently available, the average person feels insecure about making color selections. Using the method described in this chapter, you will always arrive at fail-safe color selection.

CHAPTER 6

COLOR: CURE OR QUACKERY?

The notion that light and color have healing properties was discovered more than two thousand years ago. Since that time, the supporters of this theory subject themselves to scrutiny and disdain by established medical groups. Ancient history records numerous references to color and its use in treating illness. Avicenna, an Arabian, believed in the use of color for diagnosis in the disease and disorders of his patients. He developed a very unusual chart relating to color in both the physical, as well as mental, conditions of the body. He also believed color was not only the sign of affliction, but might also be the cure for it.

Earlier civilizations recognized the importance of color, which is the optical element of light. They understood its importance in maintaining and restoring health. Ancient man built temples to worship the sun, and from the sun came light. Light is at the foundation of color. It is said that in the lost civilization of Atlantis, a person that was ill was placed in a healing room in a temple. The room was built to let in sun beams that shone through crystals, diffusing the light into colors on the patients body, in order to correct the nature of his illness.

Ancient temples built to light could be found in numerous countries across the continent. These had special rooms for those working in medicine, to heal with color. The fifth century Greek mathematician and philosopher, Pythagoras, recommended the use of color medicinally. The Zoroastrians of ancient Persia practiced a form of color therapy based on the emanations of light.

Your skin is the largest organ in your body, and is able to absorb light. When light reaches the skin, it activates a substance that lies just below the surface. This substance is known as "7-Dehydrocholesterol," and it is able to detect the nutrients in light. The 7-Dehydrocholesterol responds to these nutrients by forming vitamin D. The vitamin D then goes to work to help your body metabolize calcium out of the foods you eat.

Dr. Joseph Meites did research at the University Of Michigan, and established a link between the eyes and the endocrine system. He found that the eye contains certain photoreceptors that are responsible for converting lights energy into chemical energy. These receptors seem to have nothing to do with vision. Neurochemical pathways join these cells to the pineal and pituitary glands.

With the advent of electricity in the last century, the amount of time spent outdoors in full spectrum daylighting has diminished. The absence of daylight is the absence of needed nutrients that help keep the body in balance, and healthy. It is foolish to underestimate the effect that light has on the body, even though artificial in nature. In fact, we could take artificial light and turn it into a nutritional factory for the body.

Dr. Wurtman writes, *"At least a portion of the sympathetic nervous system responds to environmental lighting via the retina."* The retina, as proven by Dr. Meites, contains nerve endings not primarily related to vision. As incoming light reacts with the nerve endings, hormones and enzymes are released into the blood stream. Allen and Cureton, two researchers, reported that athletes show increases in motor and cardiovascular fitness, decreased pulse rate, and lowered systolic blood pressure, when exercising in low levels of sunlight.

E.G. Maas found that students were more alert when studying in natural light, versus artificial light such as fluorescent lighting. The students were more relaxed, and rated their work as easier and more satisfying. Dr. Lewy and Dr. Wever of the Max Plank Institute

We dim light for intimate dinners, light produces pictures on our television sets, we read and study by it, but most of us don't relate the import of light to health and productivity.

Lighting condition has become a science. The importance of light on people in the environment is now a priority for more than just visual performance. Environmental designers concerned with humanistic surroundings are also concerned with the acuteness of perception, health, and the improved quality in work conditions. It is mandatory that full attention be paid to how light affects our lives, particularly at work and in school.

The standard fluorescent light fixture was created because it is energy efficient, and it answered the needs of an energy crisis. Unfortunately, however, this energy efficiency does not necessarily walk hand in hand with a healthy body and high productivity. There are standard problems with fluorescent fixtures. This type of fixture flickers on/off 120 times per second. This flicker is not detected consciously with the eye unless there is a bad ballast (stabilizer) in the fixture. However, the eye subliminally is affected by the flicker, causing a type of fatigue to the eye. Additionally, the standard fluorescent fixture gives off radio emissions, or wavelength frequencies that are unseen. You can prove this to yourself simply by turning on a small portable radio and holding it up to a fluorescent fixture that is turned on. What you will hear is a loud hum. The hum is created from discharges or frequencies which are moving waves.

Each frequency can be identified by a color. Different frequencies are inherent in each type of fluorescent lamp. The problem arises with the fact that there is no balance of wavelength frequencies (color) in any fluorescent lamp. Discharges and imbalanced color waves, plus a fatiguing flicker that tires the eye, are creating fatigue for workers daily. The worker could produce approximately ten percent more under correct lighting.

If you question this information, take time to visualize your day at the office. You're working at your desk under fluorescent lights. By

RED

Positive Characteristics

The color red has always been associated with action and life. Therefore, if your favorite color is red, you love life and want to live it to its fullest. You as a red personality expect life to give you intensity in everything you experience. You will actively engage in sports, rather than sit on the sidelines as a spectator. You tend to have many acquaintances, rather than tight solid friendships. You make a wonderful "running buddy". An impulsive optimist, driven by the need to achieve success and constantly build self-esteem, would be a way to quickly describe a red type personality. Always on the "go" and striving, you, a red personality finally exhaust yourself.

Negative Characteristics

When life is not happy and exciting, you become upset and restless. If there are problems in your life, somehow they are never your problems, but caused by others. Objectivity is not your long suit. You form opinions quickly, and just as quickly change them. When life does not seem as full as you'd like, you fall into depressions and loudly let everyone know that life is not treating you well.

Burgundy - Wine - Maroon

Positive Characteristics

Burgundy, wine, and maroon are shades of red; therefore, somewhere under your exterior, buried deeply, is a red personality. You who favor one of these colors, because of life's tribulations, have come to an understanding of yourself and life. You have smoothed the rough edges. Fair, generous, social, and flexible are descriptive terms for you.

direction of their products, as well as their color. Changes are based on a change in the age group that will purchase a particular product. As a population ages, it goes through changes in perceptions. Perception changes are created by acquiring education, aging, changes in social groups, and gained economic status. A higher education will generally produce a higher income, thereby creating a social group change (socio-economics).

It is the job of color to capture your perception about a product. Certainly you wouldn't want to use a loud, vibrant color in the packaging of a fabric softener, or for a soap that is linked to fine woolens or washable silks. Soft fine goods need a color message that says, *"I will treat your fabric gently."* The basic colors of pink and blue stress softness, and gentleness. These colors also lend themselves to high volume in a supermarket. Next time you are in the supermarket, please note the color of fabric softeners. Ivory soap, which advertises its purity and gentleness is packaged in blue and white. Woolite, gentle to woolens, is packaged in blue and white.

Light pastel colors, such as pink, yellow and blue create feminine interest. The male will react to a minimal use of color, and muted shades of grey, dark blue, and strong black. The use of gold and silver, deep blue-green, deep green, and burgundy suggests quality and sophistication. Several years back, companies began to package butter in silver and gold to convey a message of a higher grade butter. Silver and gold printing on labels for wine, bags, cosmetics, and candies creates the impression of opulence.

Warm or hot colors, such as red, red-orange, orange, hot pink, and yellow on signs tell of danger or risk; but in packaging, its symbolic message is one of excitement. Conversely, cool colors like an icy blue or mint green convey a message of coolness. Blue is an appropriate color for water, water sports, and snowy outdoor pursuits. Mint green appropriately used would sell mint candy. Deepen the green for a different type of product and you will hit an upscale market.

BIBLIOGRAPHY

An Introduction to Medical Radiesthasia and Radionics
Wethered, Vernon C.W. Daniel Ltd. England 1977

Alive With Color Leatrice Eiseman
Acropolis Books Washington, D.C. 1983

Beyond the Color Explosion Renae Knapp
Rainy Day Publishing Bellevue, WA 1985

Color and Human Response Faber Birren
Van Nostrand Reinhold Company, Inc. New York 1978

Color and Light in Man Made Environments Mahnke & Mahnke
Van Nostrand Reinhold Company New York 1987

Color and Personality Audrey Kargere
Samuel Weiser, Inc. York Beach, Maine 1949

Color and the Edgar Cayce Reading Roger Lewis
Association for Research and Enlightenment Virginia Beach, VA

Color - A Survey in Words, and Pictures Faber Birren
Citadel Press Secaucus, New Jersey 1963

Color and the Structural Sense William Charles-Libby
Prentise Hall, Englewood Cliffs New Jersey 1974

Color Essence and Logic Rolf G. Kuehne
Van Nostrand Reinhold Company New York 1983

Color For Interiors Faber Birren
Whitney Library of Design New York

Colour Healing Mary Anderson
The Chaucer Limited Great Britain 1979

Color Magic Frank Don
Destiny Books New York 1977

Color Manual For Artists Arthur L. Guptill
Van Nostrand Reinhold Company New York 1962

Color Observed Enid Verity
Van Nostrand Reinhold Company New York 1980

Color Psychology and Color Therapy Faber Birren
Citadel Press New Jersey 1980

SPECIAL NOTICE TO THE READER

This book discusses various psychological personality characteristics, both positive and negative, as well as the healing properties of light and color, it is not meant to give specific recommendations or advice for treatment psychologically or physically. This book is an examination of light and color and their effect on the body and is not intended to suggest that the reader use the information within as a replacement for conventional medical approaches. It is suggested that the reader seek the expertise of a trained heath-care practitioner, psychologist or physician for diagnosis, treatment and guidance in these areas.

Barbara Colby, ASID

CHROMA PRODUCTIONS®
ISBN 0-9628138-0-X

DEDICATION

For Suzanne.
Her unfailing support, encouragement, good humor, and friendship nurtured me through the difficult times.

ACKNOWLEDGMENT

I am indebted to the many authors of books and documents, who made fact and theory available. A particular thanks to Faber Birren, whose books originally piqued my interest and drew me into a life long study.

In addition, my thanks goes to Dr. John N. Ott, an authority and author of many books on the physical affects of light and color on the body. Dr. Ott's friendship over the last several years, plus his unselfish sharing of documented information made my task easier.

I wish to thank Robert Calabrese and Ameritone Paint Corporation, for their assistance, color samples, and permission to reproduce and use The Color Key Program International Reference.

A very special thank you to Joan Solar for her support, friendship and help.

Barbara Colby, ASID

INTRODUCTION

Color & Light: Influences And Impact is an overview of how light and color relate to shaping our lives. From the time we are infants we begin to perceive color a certain way. As children we are dazzled by the phenomenon of the color in rainbows, in oil spots on the cement, and color reflection in water. Our language is peppered with color phrases, doctors diagnose by color, the color of a traffic light directs us, and factories use color to distinguish the contents of a pipe.

Color is so much a part of the world we live in that I cannot imagine life without it. Color is used to sell products, move people in and out of a restaurant quickly, and cut-back on accidents. It can create illusion; energize or sedate us; be aesthetically pleasing; or create emotion, mood and dimension.

Individual perceptions where color is concerned, come from attitudes and culture. As segmented groups change their attitudes, color trends change along with them. Politics, the economy, cultural exchanges, global influences, and the media all impact our individual perceptions, and therefore our attitudes; in turn, they then affect color choices.

This book is based on long years of research in the field of light and color and their psycho-physiological effects. It is impossible in a book this short to cover every aspect of light and color since they are both broad subjects in themselves.

The information herein is meant to inform, entertain, and stimulate the reader to investigate further the areas that they find most interesting and useful in their personal, as well as, business life.

CONTENTS

CHAPTER 1

THE AFFECT OF LIGHT AND COLOR ON THE QUALITY OF LIFE

All civilizations from the beginning of time worshipped the sun, and from the sun came light and color. In the dim recesses of man's past, some primitive native must have felt the oppressive heat of the sun on his back, turned to squint at the blazing ball high in the mid-day sky, and with a leap of imagination understood that the fiery orb far above meant survival.

From primitive times until now, light and color have fascinated people, but the average person doesn't give much thought to light — other than comfort. We recognize the discomfort when a total black-out exists, as it did one night some years back in New York; or if our computer won't work, and our TV won't play. Our conditioning has not been to think of light as visible energy, or even as having an affect on the quality of life.

Today, although we no longer consider the sun divine, we continue to pay homage to it in various ways. You can see a type of sun worship on any beach on a warm day. Because we are surrounded by so much color from birth on, unless we find we are portraying the wrong image with the colors we wear, we don't think of color as impacting our lives. The fact that our language is peppered with phrases reflecting color, tells us the important role we ascribe to it in our lives. We consistently hear phrases like —

I see the true success of man as not being measured by his economic prowess, nor his industrial genius, but rather by the extent to which he has achieved an optimal quality of life. In a general sense, an optimal quality of life can be measured by man's mastery over those areas, both personal and impersonal, which serve to influence and enhance day-to-day living. Important among these is the mastery of his environment, his drive to bend the forces of nature and to make these forces useful to him in his quest to achieve an optimal quality of life.

Light, perhaps more than any other aspect of our environment, has held center stage for centuries, as man through the ages has sought to understand and harness this phenomenon. From the earliest of times, the world has been inundated with theories and postulations as to the source of light. Does it travel? If, indeed it is radiant energy, what are its effects on people in their environment?

As we scan the history of light, we find that efforts to grapple with this awesome force of nature seems to bridge the span from religious phenomenon to our present day knowledge of electromagnetic energy. In early times, light was considered a benevolent entity, sent by the Gods who were pleased. At other times, light was blamed for the anger of the Gods who withheld it, thus causing crop failure and local miseries.

In bold relief, we see about us the far reaching advances —— the results of understanding light; and we give ourselves a justifiable pat on the back and say, "Look how far we've come!" But, in another sense, might we also ask ourselves if this is just the beginning. Have we just begun to peel back the layers of understanding and technology where light and color are concerned? Is there still a world waiting to be discovered? Today there is a renewed interest in light and color. People flock to be educated as to their use and effect. Not since the fifties have we seen as much color used on the interiors as well as the exteriors of buildings.

In the past, lighting in a building had the primary function of enabling people to see. Today we are concerned about safety and the quality of life. Without light in enclosed spaces, design, texture and color could not be seen or appreciated. Thus, our efforts are directed to the distribution of light through the image it projects, the mood it employs, and the function it serves. By the proper use of light and color we can create optimal conditions, so that we enjoy a quality of life never before experienced.

Color and light make statements! Design and color consultants use them consistently in this manner. The White House for example is not painted white by accident. Its statement is one of inpenetrability, reinforced at night by its illumination from above and below. Lighting techniques also serve to reinforce symbolic meanings, such as a traffic light which gives instructions. The use of "functional color," however, has its aim in increasing productivity, safety and efficiency. It is concerned with practical purposes; hence its appearance is secondary.

"Color coding", or an indexing of objects by color, has a language all its own. For instance, industrial pipes are color coded for quick identification in order to tell of their contents. The color code yells loudly about industrial hazards, or identifies certain equipment. The New York Transit System at one time employed 38,000 workers, and through a color code system they found they reduced accidents 42.3% in one year, coming up with a dramatic savings of $500,000.

Environment is only one factor in enhancing our day to day lives. Implicit in seeking those optimal conditions which make for a quality life is a person's moods and feelings of well being. And while light does create space, it can also be a factor in influencing how people feel. In fact, light and color can promote any feeling. Knowing this, stage lighting designers use both light and color to reinforce the feeling of a character, mood, and action on stage.

Light and color can also affect how we relate to each other, how comfortable we feel, and how well we work. The medical and scientific community has become especially interested in this. In 1984, the First International Conference on the Medical And Biological Effects Of Light was held in New York, and there is scarcely a week that goes by that the news media or trade journals do not carry articles of interest on the subject. Studies are being done on an ongoing basis in fields as diverse as psychology, architecture and ergonomics in outer space.

As a merchandising tool, color has great value in the marketing of products. We know, for example, that people have certain built-in responses that are conditioned from birth. A professional will use this information when working on the packaging and marketing of a product. Ninety-three percent of any sale made on any product is based on the packaging, and sixty percent of that ninety-three percent is based on the coloring. It's no accident that the myriad of colors you see on store shelves were put on the packages after industry spent millions of dollars doing market research. Your response to the color in packaging gives an industry leverage to capture your disposable dollars.

Light and color have a strong impact on how we feel and what we buy, but they also affect the way we behave. In a research project at Plymouth Polytechnic University, British psychologists Gary Stark, Daniel Saunders and Phillip Wookey found a difference in risk taking behavior. In their studies they found that those who sat under red lights gambled more and selected "riskier" bets than those who sat under blue lights. These findings are consistent with other research which shows that the color red brings forth aggression and competitiveness, while blue brings on calmness and security.

Our knowledge of the effects of light and color on the human body is advancing by leaps and bounds. Photobiology or phototherapy is now an established fact in the medical community. The use of infra-red or ultra-violet light in hospitals is commonplace. Full spectrum light is used to treat acute chronic depression for Seasonal Affect Disorder (SAD) patients, as well as jet lag. Soviet scientists have tested the use of full spectrum light on school

CHAPTER 2

THE FUTURE IS NO ACCIDENT

New color trends appear every year. We see them in fashion, the home, offices, packaging and automobiles. The marketing of products through color became so important that in November of 1962, The Color Marketing Group Of The United States (CMG) was founded. It spawned a professional group of people who primarily work in color and at the difficult task of forecasting colors in workshops twice a year.

CMG is composed of an eclectic group of people from every industry imaginable. Meeting at national conferences twice a year, they become a color selection think-tank. Coming from the automotive, communications/information, home furnishings, utility home, paint, and fashion fields, they focus on marketing issues, and monitor technical fields. The workshops concentrate on demographics, national events, political and economic effects, and on lifestyle, finally arriving at what they believe will be the prevailing trends affecting colors.

As Father Time brings in a new year, significant trends emerge. With the beginning of a new decade, we will see an unparalleled time of technological innovation, new materials, interest in classicism, a renewed interest in the arts and historic restoration. Additionally, a high influx of different cultures moving into the United States will impact not only businesses and lifestyles, but color and its application.

The future is no accident and is predicted by the public's move in every direction. Each time we turn on our television set and position the dial, we set an unconscious trend. When the long running television series of "Dallas" and "Dynasty" remained popular over a long period of time, women's clothes became more glitzy and opulent, helping to change colors along with the change in fashion. When we walk into a trendy chic new restaurant for the first time, we make it popular. When we go back time and time again, we set a trend. This then helps predict the style of cooking that will become the rage of a particular season. When a new President comes into office, the power behind the throne (the first lady) is a predictor of the type of social entertaining that will be done (formal or informal), style of clothing selected, and even color that becomes popular.

For instance, when President Reagan came into office, entertaining became more formal than during the time of his predecessor, and Mrs. Reagan loved the color red. We the public, felt the impact of her color choice, as Reagan Red began to appear everywhere. When the art exhibit "A Day In The Country" went across the United States in 1984 and was viewed by thousands of people, suddenly everyone began to ask for the French Impressionists' colors shown in the art work, and we had the mauveing of America in 1985.

If we who forecast trends see into the future by logical steps, the public, who may not make the steps logically, still tend to feel them unconsciously. The forecasting of colors is not a "crap shoot", but based on meaningful information, moved into different marketplaces to meet the individual tastes of the consumer in that specific marketplace. Color families are chosen, and then each individual marketplace will tone the color to be useful for its consumer area. Certainly, colors used in a kitchen on canisters will be bolder than a color used in a carpet for the floor.

Automotive manufacturers also follow color trends and will use colors coming from a particular palette for a future year, but will create a color that is based on the model automobile to be purchased. Automotive manufacturers are aware that women are more apt to be interested in color and convenience of handling, whereas a man regards his automobile as an extension of himself. Enter color! The automobile manufacturer realizes that those making a purchase of the magnitude of an automobile do not want to worry about criticism from their peer group, so the manufacturer always uses quiet colors in upholstery. This leaves an easy decision for the purchaser.

As for exterior color for automotive selection, the manufacturer determines the character of the automobile to see if it is a sports car, a luxury auto, or an economy style. Based on the type of person who will purchase a certain model style, colors are chosen. The colors then follow the ascending trends. Product development departments know that a person purchasing a sports model is attracted to sleek sensuous colors like red, black and yellow, whereas the luxury car buyer will purchase deeper, richer colors in metallics and high socio-economic colors. The economy purchaser tends toward basic colors without metallics, like beige, white or blue. As trends change, these same colors are lightened, brightened, or are more saturated, and then tinted into various stylish tones of the times.

When realities change globally or here in our country, our lives and perceptions are impacted, they create attitudes. Attitudes bring about responses to external forces. These responses work to develop opinions, which are at the base of an individual's decision making process toward color, design, and product choices.

Vast currents of change and fundamental evolution from both political and economic upheavals influence change. Our fast burgeoning shift in technology and the media, which brings the world into our living rooms, influences our senses. Our congested highways and overcrowded cities relate to our homes becoming increasingly important, which in turn tells us this is a place people will go to for comfort and privacy. If this is true, then a new division of space in the home will be required. So space, privacy and flexibility in the home will be key issues around which homes will be designed in the future.

Following logical steps, we know that this means that the kitchen and master bedroom/bath areas will be our fastest growing spaces in homes. The kitchen will be at the top of the remodeling list, with the master bedroom/bath area following a close second.

We know our population is aging. By the year 2012, those fifty years and older will control fifty percent of all the spending power in this country. This means we're going to have the graying of America, and wrinkles will be trendy. It also means that this demographic group will be looking at making residence changes, buying second homes, and enjoying leisure travel. Enter a trend in travel and a big trend toward cruise vacations. Cruise trips should increase four-fold by the year 2012.

Demographics of tomorrow communicate a message of rapid urbanization. Industrialization and colonization of outer space with its zero gravity will bring dazzling new manufacturing techniques to we here on Earth. As technology moves ahead and more properties are available that create color, new lusters and iridescence will appear for each individual marketplace. We'll be surrounded by paints that change color or have a chameleon effect. Stocked product lines will become smaller as manufacturing becomes more capable of producing custom colors from new technology. These new technologies will create pressure and issues for manufacturers.

The future is no accident; it comes from condensing what you see, hear, and read. Color trends come from the foundation of changing attitudes, needs, demands, the economy, global barriers being lowered, influences of cultural offerings in the marketplace, and lifestyle changes. The age, purchasing power, and attitudes of a demographic group apprise we who forecast, of basic buying preferences. Mass communication, global/cultural influences, politics, the economy, and national events determine issues from which to forecast trends. In other words, the issues create a bridge to ascending trends. Certain colors are then related to those perceptions; then a new color palette emerges on the American scene.

CHAPTER 3

THE FUNCTION OF COLOR

How Color Functions Commercially

The function of color in hospitals, schools, offices, retail businesses and industry is not a matter of aesthetics, but is concerned with more practical aspects. In this case appearance becomes secondary to more practical purposes. When a technical approach is used, and color is properly applied, it has a silent voice that yells loudly.

It is common knowledge that people are impacted by their environment both psychologically and physiologically. Color studies have shown that color increases cortical (brain) activity, and the functions of the heart, blood pressure, and respiration rate. Therefore, the colors that surround us in a commercial environment become critical to the task at hand.

Color acts to serve us in many ways. Color can delineate space and help to code an area for future reference. In a restaurant it can move people through more quickly and stimulate appetites. It marks hazardous areas, makes seeing more comfortable and efficient, identifies the contents of the material contained in pipes, creates volume, and gives a get-well message to patients in a health care facility.

Although most people don't recognize the utilitarian side of color, they are affected by it daily. Playing a vital role, color reinforces sales messages in retail establishments. Color says, "Look At Me!", and then tells about the product and price relationship.

Schools

With short attention spans and lower grade scores prevailing in the classroom today, schools need to look more closely at color and how it might be used effectively. Richard J. Michael, a San Francisco instructor found that by using slides with the predominant color of cool green during a lecture, and testing immediately afterward, students' test scores rose as much as forty percent. Red came in second, and the color blue lagged in third place.

Fatigue and eye strain are common to students. Much of this is due to incorrect fluorescent lighting and glare zones, which constrict the pupil of the eye. The students over work their eyes because of the glare. Having taught interior design on both a high school and college level, I always find it amazing when I walk into a classroom and find drafting tables with white laminate tops. The glare of fluorescent ceiling lights reflecting on the white tops creates an unbearable glare for a student's eyes.

Brightness ratios from furniture, walls, equipment, etc., need to be uniform and without glare. Colors that give good light reflectance of at least fifty percent, and do not visually distract the students from their studies, are appropriate colors to use. Light reflectance values for colors can be found in the back of a Munsell Color Notation manual. Mid to light color values give the best reflectance.

Students at different age levels tend to act differently. A child on a kindergarten, first or second grade level will be far more animated and lively than a student in sixth grade; or high school. Since colors adapt well to activity levels, before selecting a color, the grade level and age of students should be taken into consideration. Bright warm colors tend to divert and are outward directed colors, but so are kindergarten and early elementary grade students. Use bright warm colors on doors, cabinets, and then soften down one of the colors, or use a more neutral warm beige or gray on side walls.

Ceilings should always be white for reflectance qualities, but adding one-half gallon of the side wall color to every five gallons of ceiling white tends to keep the reflectance quality, but give it a soft tint. Carefully adjusting the brightness to softer colors of the same hues for the side walls avoids monotony and doesn't distract or make the eye work overtime. Yellow-based pinks, such as salmon, coral, peach, or a soft yellow-orange, will give active kindergarten and elementary school children pleasant surroundings in which to learn.

When students face in one direction, the wall behind them may be treated in a deeper value of the same color as the side walls, or one of the brighter colors used on cabinets or doors. This is another way to break visual monotony in a classroom. When the students catch a glimpse of another color, it is visually relaxing.

Upper school students need to concentrate and direct their attention inward. Light to mid-value cool colors, such as blue-green, teal blue, greens, warm beiges and grays are appropriate. End wall treatments behind the students may be used in a harmonious Emerald Green, deep Teal Blue, or a deeper value of a blue-green.

Athletics

Manipulating behavior with color in athletics has proven to be a winner. Hobie Alter changed their Hobie Cat catamaran image by striping their sails with bright splashy colors to give them a fast, sleek, fun look; the change dramatically increased the volume of their sales. An athletic director at the University of New Mexico decorated his teams' dressing room with red, and the dressing room of the opposing team in blue. This follows studies that show red impels people to action in an aggressive manner, while blue sedates.

Alonzo Stagg when he was head coach for the Chicago Bears had two dressing rooms for his players. One was painted blue for rest periods, the other was painted red for fight talks. Felicia Fuller, a color psychologist, believes baseballs should be a bright coral red, and golf balls any color but white. This makes sense since the eye is able to follow color more easily.

Dr. John Ott, well known color/light researcher, was called in by Rex Bowen, a Cincinnati Reds scout. The Reds were way down in record scoring for the pennant. The subject of the player's baseball caps with their green visors came up in conversation. Dr. Ott suggested that they be changed from green to grey on the undersides of the visors. Ott said it would make a difference in the performance of the players. Bowen convinced the front office to give it a try. After the Reds instituted the change of color on the caps, running speed increased, mental alertness improved, and they won the pennant! When the light reflected off the underside of the green visor, it created a neurochemical reaction through the retina of the eye. This affected the entire endocrine system, which controlled the body chemistry of each player. Subsequently other teams changed the visor color on their caps.

Restaurants

Our conditioning is so strong where color in food is concerned that the food industry uses more dye in food products than is used in textile manufacture.

Color plays a particular role in restaurants. An appetite color scale, in which the effects of color were measured, placed red-orange at the top of the appetite scale. Bright warm colors like red, yellow, and orange tend to stimulate our digestive juices, while soft cool colors tend to retard them. Studies show there are peaks and valleys when appetite levels are connected to color. The top peak is found in the red-orange area. It drops off at yellow-orange, and picks up again at yellow. The low point is in the yellow-green area. Appetite appeal is restored in cool-green and blue-green, and shows a definite drop in the purple area. Poor colors are purple-red, violet, yellow-green, green-yellow, gray, olive, or mustard.

Fast food restaurants want to increase appetites and move people through quickly, so you will note that red-orange and yellow are often found in this type of restaurant. Certainly in a more refined restaurant where mood, food, service and surroundings create an ambiance so the patron has a dining experience, color becomes only one of many elements reinforcing a design statement.

The Workplace

Human performance, and the attitude of a worker, starts with physical comfort in the work environment. Research has shown that at the top of the list of specific job selection, for most workers, is the comfort and appearance of the environment in which they will be working.

Color studies done by the National Industrial Conference Board of three hundred and fifty companies showed that 64.7 percent stated that color had improved lighting; 27.9 percent reported production increases; 30.9 percent noted an improvement in the quality of work performed; 19.1 percent commented on reduced eye strain and fatigue; 14.7 percent credited color for reduced absenteeism and better moral.

The U.S. Public Health Service and The Public Building Administration in Washington did a two year study and investigation in the field of vision, illumination and color. Three conditions were analyzed:

1. The original room
2. The room with the addition of new light fixtures
3. The room with the further addition of color

Where proper lighting and color were used, there was a 5.5 percent production improvement. In gross payroll, 5.5 percent equated to $13,229.00 divided among ninety-five employees. This means that the right illumination and correct colors would be worth $139.25 annually per average employee in American industry. So color put to the test is found to repay its investment.

Color engineers exploring the physical and psychological reactions and benefits of color concluded that the health, comfort, happiness and safety of those being tested were due to the results of the energy in color. A person subjected to a given color for as little as five minutes will change both muscular and mental activity. Tests have shown that muscular activity measures at 23 empirical units under ordinary light. It advances slightly under blue light. Green light increases it a small amount, but yellow light will raise the activity to 30 empirical units.

The perception of certain colors can change the feeling of length of time. Two groups of salesmen, whose watches had been removed were asked to guess how long a meeting had lasted. The two colors being tested were red and green. In the red room, the men guessed six hours. However, the men in the green room estimated the meeting lasted three hours.

Safety Colors

Safety colors were formulated in 1944 when Faber Birren, a long standing color expert, worked with the DuPont Company to develop a color safety code that was later accepted by The American Standards Association. These are used consistently. Every time you see a workman in the street with a bright fluorescent orange vest, or a barrier marked with yellow and black stripes, you are experiencing safety code colors that identify information for you (see Figure 15).

YELLOW

Yellow used in combination with black, is used to mark hazards where one could stumble or fall. It is painted on barriers, dead-ends, low hanging beams, or edges of platforms. A bright yellow entrance or exit bumpers at doors caution vehicles to steer clear. Since yellow is a high visibility color, it is used to mark aisles, and to paint on overhead moving objects, such as a crane.

programs that lead to improved employee moral will generally lead to quality in products with less waste of time and materials.

Exterior Environmental Color

Not since ancient times have we seen as much color give a boost to our exterior environment. In ancient times color was used to symbolize. Archaeologists were in awe, when to their surprise they uncovered great ruins to find beautiful marble statues, great temples, and frescoes, all tinctured with color. Ancient man had smeared bright coatings over marble columns, ornaments, and sculpture.

Uncovered Greek statuary was tinctured in red, green, blue and yellow. Frieze backgrounds, relief figures on walls, and decoration were reasons for ancient cultures to use color. For them, it was natural to use color, or paint skin tones on statuary; after all didn't the human figure have this color in its skin tones?

To the ancients, architecture as well as statuary became an object of art, and as such, color became a part of the object. If one could travel back in time to view early architecture, in every country you would find brilliant colors on roofs and buildings. As the Renaissance moved in across the world, color lost out on the exterior of buildings. It was thought to be pagan. Then in 1920, we find color beginning to show up once more, on the exterior of buildings in bright colors on modern buildings in Holland.

LeCorbusier used color on projects to emphasize or to make transitions from one area to another. It wasn't until the seventies that post modernists finally brought the importance of color to the exterior of buildings in the United States on the west coast. Michael Graves, an accomplished artist in his own right, took it to New York, and a new color revolution began.

Planning a color scheme for the exterior of a building is based on how many colors will be used in a scheme, how often they are repeated, how large the area is to be covered, and whether one is trying to blend the architecture to the background (camouflage), or accent the architecture to intensify its qualities against the background (contrast). It is best to use a blueprint or an elevation that is drawn of the structure to be colored. Many overlays can be colored in on tracing paper until one arrives at an acceptable, pleasing scheme .

Wall Art

After World War II, some early environmentalists concerned with the brutalizing effect of the world, were convinced that art could be used on the overscale side of a building to create an artful environment. In the sixties wall art began to replace graffiti, and from that time forward it has produced depth and illusion in our modern environment. Originally it called attention to slum areas, and covered graffiti; now wall art brings decoration and vitality to the surfaces it is used on.

Wall art can be seen from creativity by Jean Phillipe Lenclos, a French designer, to Op artist, Carlos Cruz-Diez in South America, to America's ghetto artists with a message. Color is turning up on buildings, streets where entire floorscapes are created, on the sides of buses, and on freeway underpasses.

Color is everywhere! San Francisco is famous for her graceful Victorian "Painted Ladies". Architecture in this colorful northern California city, from turn of the century times, looks better today, with colorful painted faces, than when the buildings were first built. Travel across the world to Turin, Italy, and you will find a color program for the city that was discontinued in 1854 now being restored. All buildings, major routes, and secondary arteries are color coded, a wonderful cueing mechanism for finding ones way around. Industrial sites in France, Japan, Italy and Brazil are being turned into colorful celebrations. Silos, storage tanks, warehouses, industrial machinery and walls, create color impacts along roadways, a prescription for blah environments.

Whether here in America, or in any foreign country where you travel, color has come home to celebrate life in a dynamic way.

The original information for color harmonizing that never fails was discovered by a man named Robert Dorr. While working as a poster painter in 1928 for a large chain of theaters, he became aware of the fact that all colors contained an undertone mixture of either blue or yellow.

He noted, as he painted in the pigment on a star's face on individual posters, that there was a pigment relationship that varied for each person. In some cases, some skin pigmentation would be peachy, while for others it would be pink or rosey. He found that the skin pigmentation could be divided into two categories, which he defined as Key 1® Colors or Key 2® Colors. The Key 1® Colors have a blue undertone mixture, and the Key 2® Colors have a yellow undertone in their mixture.

It was in the 1930's that Dorr used this differential and successfully formulated the first correlated textile group for a fabric manufacturer. After that time, he became well known as a professional color consultant who worked for cosmetic, automotive, advertising and graphic companies. Mr. Dorr patented his system, eventually retired, and sold the system, to the Ameritone Paint Corporation, a subsidiary of Grow Industries. The system registered under the name of "Color Key Program International Color Reference®," is now represented throughout the United States by Ameritone Paint Corporation in Compton, California, and Devoe & Raynolds Paint Corporation based in Louisville, Kentucky. Each of these paint companies has produced its color selector guide (paint decks) based on this system.

The Color Key Program® consists of the two color palettes of Key 1® Colors and Key 2® Colors. Key 1® Colors have a blue pigment base. Key 2® Colors have a yellow pigment base. Ameritone Paint Corporation and Devoe & Raynolds Paint Corporation, have arranged each of their color selector guides (paint decks), in Key 1® and Key 2® colors so that the public can easily choose colors from their paint chips that harmonize. Mixing and matching colors from one of these two individual groups will always produce fail-safe color harmony.

This system has proven to be a fast effective method for those who work consistently with color, whether an architect, interior designer, or image consultant. Using this method, the average consumer can make fool-proof color choices that always harmonize. To give you an example, shown in Figure 1 is a neutral beige. Next to it is its first cousin, brown. The two together just don't feel harmonious. Why? Beige is a neutral and should mix well with brown. Yet the beige has a pinkish cast to it, and the brown tends to look yellowish. What happened?

What happened is a clash of the underbase mixtures. You can take any basic color, such as red, blue, yellow, green, etc., and change not only the color, but the look and feel of that color, simply by changing the underbase mixture. What happened in my illustration is that blue-based beige is not harmonious with yellow based-brown, causing dissonance.

Now look at the colors of beige and brown in Figure 2. You see total harmony, side by side, in this example of a blue-based beige next to a blue-based brown.

In Figure 3, note how a yellow-based beige looks with a yellow-based brown. In Figure 4, you can see the two different beiges in the two different undertone mixtures. It is very clear that the beige to the left with a blue-base has a pink cast. The beige shown on the right side has a yellow-base and a decided yellow cast.

Now look at the supposed neutral grey in Figure 5. The grey on the left has a blue tint, while the grey on the right has a totally different cast.

White, an assumed neutral, in Figure 6, shows a blue-based white on the left, which is a snowy white. The yellow-based white on the right is a creamy white. Try to put these two whites together, and you will have dissonance.

Obviously there are no universal neutrals in dealing with color harmony. Each defined neutral belongs to its own grouping, in either a blue-base Key 1®, or yellow-base Key 2® category. Each one of the color chips in the Color Key System® has a number specifying the mixture guide; this guide begins with the figure of either one, or two, such as 1UB1B, or 2UB1B. The beginning figure specifies either a Key 1® Color, or a Key 2® Color as shown in Figure 7.

Although I have illustrated basic neutral colors, the Color Key System® applies to every color available, as well as to human beings with their physical differences. Although we are not all created equal, we are created uniquely. Basic pigmentation of the eyes, skin and hair requires that our uniqueness fall into one of the Color Key Palettes®. Hair may change color over a lifetime, due to age or dye. Your eyes may change due to a choice in contact lenses, but the one constant is the skin pigmentation. When hair turns grey, or our bright blond tresses darken down, they do so in the same color grouping or consistency in pigmentation.

Therefore, people who have Key 1® pigmentation will have rosier complexions and silver grey hair; whereas Key 2® pigmentation lends itself to a peach complexion and a creamy white type of grey in hair coloring. Key 1® blonds tend to be platinum or smokey, while Key 2® blonds are a golden or honey tone. Key 1® redheads have an auburn look to their tresses, and Key 2® redheads sport carrot tops. You'll find raven-colored black hair on Key 1 people and a deep charcoal hair coloring on Key 2® people. Brown hair on Key 1® people will tend to have reddish highlights, while on Key 2® people the hair will be a chestnut color. See Figure 8.

I had a student in one of my college classes that came to me one day and asked about her hair color, which was a golden shade. She was a Key 1® type in her late fifties. I explained as diplomatically as possible that she was dying her hair in a Key 2® color and wearing Key 2® colors, when she should be dying it in a Key 1® platinum and wearing Key 1® colors. The next week she came to class early. She was wearing Key 1® colors and had her hairdresser change the color of her hair. I quietly congratulated her on the new effect. It had taken years off her age, and her complexion glowed.

As the other students filed into class, they all immediately noticed the new look, cooing over the effect. The student who had made the changes waited until all the other students left at the end of class, then came up to me, hugged me and said, *"This segment of your classes was worth the whole course."* I don't think her feet touched the ground that night.

It is important in choosing clothing colors that you remain in colors that flatter and bring out the highlights in the pigmentation of your skin. At the end of this chapter, I have illustrated some of the basic colors in Figure 9, in both a Key 1® and Key 2® palette to give you some guidance.

Whether you are Asian, Latin, Black, or Anglo-Saxon, your basic pigmentation remains consistent. It is so consistent that even when you suntan, your tan only deepens the pigmentation you were born with. For this primary reason, if you are a Key 1® type, "quick tanning" lotions do not work well. Yes, you will get a tan, but it won't be the deep bronzed tan you are seeking. This type of lotion tends to have yellow base undertones, and therefore is best used by Key 2® types.

If you are Asian, Latin, or Black, although your skin may appear as one color to those outside your culture, your skin is also based on the same medical evidence of differences in skin pigmentation. Therefore, any cultural skin type will fall into one of the Color Key Palettes.

There are times when color combinations are used to achieve a dramatic affect in fabrics, or interior design in commercial establishments. Colors are taken from Both Key 1® and Key 2® groups and used successfully in combination. This is done by a professional who is adept at combining colors from both groups. Using colors in combination from both groups together is referred to as "Cross Keying.®"

There is one color in each Key that you will not find in the palette of the other Key. Magenta, a Key 1® color, will never be found in the Key 2® palette, since when you add a yellow underbase to this color it becomes a yellow red. Orange, a Key 2® color, will never be found in the Key 1® palette because when you add a blue base to orange, it becomes a muddy color.

Using the Color Key Matching Reference System®, whether in making your surroundings more flattering, or helping you select and understand personal preferences, I encourage you to use colors you may never have used. Even if new to your experience, they will make an exciting difference in your life.

Further information on the Color Key matching system can be received by contacting:

Western U.S.

Ameritone Paint Corp.

18414 S. Santa Fe Avenue

Compton, California 90221

Eastern U.S.

Devoe & Raynolds Paint Corp.

4000 Du Pont Circle

Louisville, Kentucky 40207

CHAPTER 5

THE BODY'S COLORED FINGERPRINT

Our bodies are composed of atoms containing electrical charges: positive and negative, electrons and protons. It is assumed that this atomic action is directly related to what scientists call the "electromagnetic energy field." This energy field is created by electrical charges that emanate from inside the body to the outside of the body, and are the unmistakable result of the action of the body's electrons and protons. The electromagnetic energy field can be seen on a Kirlian photograph in differing intensities, colors, and flare patterns.

Before the Kirlian camera and discharge photography were discovered by electrical engineers in the Soviet Union, people in metaphysics called this vibrating field the "aura", or "corona." Discharge photography was discovered many years before the Kirlians became interested, but was used to detect flaws in metal used in engineering. Semyon and Valentina Kirlian, a husband and wife electrical engineering team in Russia, became interested in the effects that discharge photography could reveal about the electromagnetic energy field. They went to work systematically with living organisms to look at the electrical energy field that was not visible to the human eye. It was only after the Kirlian photograph came into being that the electromagnetic energy field became an acceptable scientific fact.

In Kirlian photography, an object is placed on top of a flat metal plate on which a piece of film has been positioned. High voltage electricity is pulsed through the metal plate at a very low amperage. The electricity passes through the film, exposing it, subsequently producing an outline of the object, plus patterns, flares, and colors when color sensitive film is used.

In the 1970's, Thelma Moss, then working at UCLA's Neuropsychiatric Institute, became interested in the Kirlians' research and traveled to the Soviet Union to study what the Kirlians had discovered. Returning, she compiled thousands of films on plants, animals, and human interaction. It is due to her interest and research that work with Kirlian films was brought to the attention of the American public.

In the early 1900's, Dr. Oscar Brunler, a physician and physicist proved that every organ in the body, in a healthy state, has a specific, measurable, vibrating wavelength frequency. Pictorial evidence in ancient cave paintings in many countries shows a halo of light around the body, head, and hands of figures depicted in the cave art. I believe this to be ancient man's way of recording the electromagnetic energy field for all of posterity to see. The electromagnetic energy field is at the base of so called natural healing, or a "laying on of hands." Therapeutic touch is the transference of highly excited electrical energy produced by the body, from one human being to another. Written history, which goes back 5000 years, states there were "natural healers," and in early France and England, those with this type of touch were said to have "the king's touch."

Dolores Krieger, who has a doctorate in nursing, teaches "therapeutic touch" to other nurses, and the public in general. Hospitals around the country use therapeutic touch teams for a relaxation effect before the administration of anesthesia. It is used in cardiac units prior to insertion of a pacemaker, or in emergency rooms on patients in pain while they await needed medical attention.

If your body is able to transfer electrical energy, there must be a way for it to be conducted. In other words, there must be a field of called, “the phantom leaf affect.” It is called that because although the top of a leaf is amputated, it is still visible on the leaf, when that leaf appears on a Kirlian photograph. What you see on the film is invisible to the eye. The leaf’s alive sustained energy is able to be captured on the Kirlian photograph. This invisible electrical energy may be the reason that an itch is felt on a severed limb. Although modern medicine would never accept this theory, could these electrical vibrations, which never die, tell us that a person is experiencing a phantom effect from the invisible electrical energy on a limb that no longer exists?

The electromagnetic energy field has been observed, by those with the unusual powers to see them, as being as small as 1/4 of an inch away from the body, while others will emanate up to one-foot away. Numerous studies show that a person’s electrical field will change in size, shape, color, and flare patterns as a result of disease or disorder. Either good or bad emotions also exercise the same type of influence. Thousands of films have been taken since Kirlian photography was introduced into the United States. The colors seen on these photographs are blue and white, red, and red-yellow or red-orange. No specific meanings have been established or attached to certain colors where Kirlian photography is concerned. Studies do show that tension, anxiety, and emotional excitement will show up as red, or a variation of red. The electrical field of a healthy human subject will emanate the color of blue-white.

Subjects who were drunk when their fingertips were photographed showed a decidedly rose color. Smokers showed a changed pattern and color as opposed to those who were non-smokers.

Scientific evidence through photographs is just too strong to ignore. Our body’s colored fingerprint silently makes statements about both our physical and emotional make-up.

If it were possible to enlarge the Kirlian Camera to full body size, such as an X-ray machine, the assistance in medical diagnosis would present unprecedented promise.

Hippocrates, the father of medicine, practiced his profession on the island of Cos, in the Greek Archipelago. The great temple of Aesklepia was built there, and dedicated to Aesculapius, God of the Sun (color), medicine, and music. Hippocrates, as others before him, recommended color to help the body heal and stay healthy. The ancients understood that from sun came light and that light is the foundation for color.

In 1666, Newton passed light through a prism and split the light into continuous bands of color. Newton was an atomist, and supposed that light was generated by particles which spun about and moved forward. This made Newton the real father of modern day physics, and led to a new doctrine on the refraction of light. Newton's theories on color and light are an established fact accepted today. Newton's information marked the way for the theory of light being radiant energy. This meant that light had substance. Newton postulated that if light had substance, then why not action.

With the advent of chemistry, modern medical techniques, and surgery, color lost out until the latter part of the nineteenth century. It was then that Edwin D. Babbitt's efforts brought color therapy to the forefront once more. Through Babbitt's teaching and lecturing, color practitioners began to spring up everywhere. Babbitt was part scientist, part mystic, and part physician. He had an interest in ancient philosophy, a cosmic interest in life, as well as an interest in modern science. He was enthusiastic, believed in his theories, and because of his enthusiasm and strong belief system, along with logic, he was able to create public interest.

In 1878 that he published a book, *The Principles Of Light And Color* which was praised by some, and denounced, as well as attacked, by others. The book was on color therapy. The uproar over his book only brought more attention to his theories and color therapy.

He saw in light an all encompassing harmonic law, a unity between electrical action, as well as a potency, when light was used with color. He believed that the colors of light were a lower form of higher vibrational energies. He felt that the vibrational characteristics of color, when used with light, had different chemical, and therefore therapeutic, properties.

Babbitt knew that the human body is composed of cells and that cells are the basis of all life. He also knew that the body's cells could not survive alone and needed help in the form of nourishment from within the body. The nourishment came from inorganic substances. Today these inorganic substances are known to us by names such as Potassium, Calcium, Iron, Phosphorus, etc. These are the substances that work to help the cells undergo changes, so that life within the body can be continued.

Each substance is made up of atoms. The atoms contain electrical charges, both positive and negative, and are called electrons and protons. Babbitt's whole theory is based on the atoms within the body and their movement and interaction. Babbitt assumed that when something in the body happens to throw off the charge of the electrons and protons, the body would experience a form of short circuit in its electrical system and cells. Hence disease or disorder would ensue.

Following on the heels of Babbitt was Dinshah P. Ghadiali, a medical practitioner from India. He moved to America in 1908 and later became an American citizen. It was in 1897, in India, that the niece of a friend was dying from mucous colitis. Accepted drugs of the day were not working. Dinshah had read of Babbitt's work. Using his information, and an indigo glass bottle as a colored filter, plus a kerosene lamp as light, he cured the woman. This was the inception of Dinshah's work as a color therapist.

In 1920 he gave his first lecture on what he called Spectro-Chrome, in New York city. Dinshah believed that light and color worked through specific glands, cells, and the blood circulatory system. He took Babbitt's information, refined it, and then announced to the world his perfected method of color for healing.

Babbitt had worked with a large pane of glass called a Chromalume. The glass had different sections of colored glass through which daylight passed to heal a person. Dinshah perfected a heavy iron lamp, used specifically with twelve different colored glass filters. Light was shown through the colored glass filter directly onto a nude area of the body, or the nude body as a whole.

Dinshah's theories were based on the premise that our bodies contain many chemical substances, and these substances contain a certain balance of color waves, or wavelength frequencies. He felt that a disorder in the body constituted an imbalance in the wavelength frequency of the chemical substances. Radiating the body with light and color, he postulated, stimulated the body's immune system into action, thereby allowing it to heal itself in a normal fashion.

At this point the reader may ask, "Does color have healing properties, or is this just quackery?" We can begin by examining a procedure that is acceptable in every hospital's neo-natal unit, that of bilirubin lights. These are used to treat newborn infants with underdeveloped livers. Formerly babies with hyperbilirubinemia died. As time went on, hospitals gave blood exchanges to the babies. Still many of the babies died. Those who didn't die were physically or mentally retarded.

One day in a lab, accidentally, tubes of blood that were waiting to be checked were left on a window sill. When the lab technician checked the blood, they found that the bilirubin level was lowered, due to the blue side of light. Subsequently a light box was produced with a canopy of blue lights. After treating a baby in this light box, the infant was able to be sent home healthy and happy. That ended the blood exchanges. The blue side of light accelerated the conversion of the breakdown of bilirubin into bile, proving that the blue side of light had a direct effect on the liver and the kidneys.

Recently, a group of biochemists from North State Texas University reported that they had isolated an enzyme in a cell. This enzyme was named Triosephate Isomerase; shortened it is referred to as TPI. It is a crucial enzyme that seems to stop working as the body grows older. TPI stimulates white blood cells to fight infection in the young, but in the elderly, according to Robert Gracy, head of the Texas team, TPI is chemically altered. For some unknown reason it takes on a negative charge and becomes totally useless. It lingers in the aging cells, but it doesn't do its job. The immune system fails, and disease sets in. Since light and color directed on the body seem to recharge molecular activity, is it possible that a positive charge could be placed on the enzyme, with light and color, thereby creating an active immune system?

Molecular activity is directly related to the electromagnetic energy field. Color comes from breaking down light. Light slows down and forms electrons, and this is what the body is made of in differing degrees and densities.

Although modern medicine puts no stock in color therapy, we find the medical community using color in the diagnosing of illness. Diagnosing by color ranks next in importance to taking a pulse, blood pressure, and temperature. Color symptoms are well known to doctors and are an indicator of many illnesses or ailments. Stop and analyze for one moment your last visit to your doctor. The procedure generally entails an interview on your symptoms, a check of your temperature, pulse, the color variables in your eyes, your throat, the skin, and fingernails.

Our knowledge of light and color, and their effect on the body is advancing by leaps and bounds. Experimentation and research daily seem to fall into the category of fact. Scientists in the fields of psychology and medicine indicate that both light and color responses are total. They influence the emotions and the whole of our organism. They affect the autonomic nervous system, as evidenced by the changes in heart rate, respiration, cortical activation, and muscular tension, as well as other functions. Even though science knows this to be true, the medical profession ignores the use of color and light (color therapy) as being a viable alternative method, or a compliment to the function of medicine today.

How does color therapy work? The energy of color and light are transformed in many ways. The autonomic nervous system that controls the heart rate, lungs, intestines, glands, and other internal organs is stimulated by nerve endings in the skin. Differing wave lengths are able to penetrate into the capillaries, and are absorbed by the blood. This, in turn, creates wide spread changes in systemic reactions. The process of passing a colored filter in front of an incandescent light changes the wavelength frequency. The light acts as a vehicle, similar to a truck, carrying the color into the body.

Each color used can be equated to an element in the body through the use of a spectroscope. When elements are in an active state (ionized), each element emits a characteristic set of colored lines on a spectroscope. These lines are known as Fraunhofer Lines. These are always in the same relative positions, as far as elements and color are concerned. You might say that they are an element's identifying fingerprint. Fraunhofer Lines are seen on a spectroscope as brightly colored bands of differing intensities. However, if you pass a beam of light through an element under activation, the bright bands disappear. In their place there appear dark bands without color. Where the element had been emitting energy of a certain frequency, it is now absorbing the energy of the same frequency. The dark bands are called "absorption lines."

If it is true that an element can absorb the energy of light when the element is in a live state, does it not follow that the human body that is composed of live elements could be influenced as light and color pass into those elements? This was Dinsha's premise based on the body being made up of live elements.

Having personally witnessed, as well as experienced, healings from color and light, I choose to believe that there are benefits for the human body from them. Light and color brought total recovery from a life threatening surgery and a crippling spinal condition in 1981. Several years ago I was diagnosed as having a tumor on the right optic nerve. Surgery was suggested by the physician. I opted for color therapy. The tumor disappeared. This is down on record on a UCLA cat scan. Several years have elapsed since the day the tumor on my right optic nerve disappeared. I have had no recurrence or any symptoms.

I have witnessed the abating of a two year old decubitus ulcer on the foot of a seventy-eight year old woman. Three top surgeons in the country told she and her family that the foot would have to be amputated. After six weeks and thirteen tonations of light and color, her foot was normal (see Figure 12, 13), remaining that way until she died six years later.

In the case of Ms. J.S., a badly broken wrist was healed in three weeks to the day, and came out of the cast with 55% mobility, due to light and color therapy. This is also down on record.

Miracles? Maybe, but I like to believe it is based on the physics of light, and chemistry of the body. Working together they created a natural healing reaction. It boggles my mind that orthodox medicine is not receptive to color therapy or other alternative practices of healing. Since 1846, when the AMA's code of ethics was formulated, allopathy closed its door on the homeopath and alternative practices.

The rising cost of health care is making people have to take more responsibility where the health of their body is concerned. For the same reason, in the not too distant future, Doctors will have to learn to practice medicine differently, and will have to take a look at the benefits alternative practitioners could provide for their patients. Many viable methods that do not fit into mainstream medicine are working and helping patients that the medical community cannot seem to assist.

Albert Einstein said, *"It is possible there exists human emanations which are still unknown to us. Do you remember how electrical currents and unseen waves were laughed at? The knowledge of man is still in its infancy."*

Color therapy is not a swift method of healing, but it is a natural method that slowly rebuilds and regenerates the body.

Some people would place it in the category of fringe medicine, while others do not accept it at all and consider it quackery — or a way to delude the public. In answer to those who doubt, I pose this question; Do you remember not too many years ago when owning acupuncture needles in this country was a crime, and acupuncture was viewed in the same light?

If we continue to peel back the layers of understanding, we will find that there is still a world waiting to be discovered where light, color, and technology are concerned. Then the doctor and the patient will stand on the threshold of tomorrow today.

CHAPTER 7

THE NUTRITION OF LIGHT

Light as an energy source is vital and has more effect on the body indoors, as well as outdoors, than any other source of energy. Today, there is a renewed interest in light and its effect on the body. This is due to the media's interest in Seasonal Affect Disorder (SAD). SAD people become depressed, irritable, anxious, socially withdrawn and sleepy as the days grow shorter with the onset of fall.

Several years back, psychologists working on this problem discovered that by exposing patients with SAD to full spectrum light, the patients' symptoms dissipated, and they were able to lead a normal life. Data also demonstrates that more suicides and depressions occur in the winter months when days are shorter, than in spring and summer months when days become longer and more daylighting is available. Mood disorders in animals who hibernate reinforce these findings. This information, along with other research and our New Age thinking, has brought us to the doorstep of light playing an important role in our lives. Whether physiologically or psychologically, Mother Nature has irrevocably tied the psyche of a person to his/her corporal body.

Light is being looked at as a form of nutrition. Top doctors are writing about their promising results using light in controlled studies to bring about health. Even draft board figures seem to corroborate findings on light's healthful effects. The figures show that the healthiest specimens come from the Pacific coast region, where people receive more direct light on their bodies for a longer period of time throughout the year.

Two Soviet scientists, N.M. Lazarev and M.V. Sokolov, at a meeting in 1967 of the International Congress On Illumination that was held in Washington, said, *"If the human skin is not exposed to solar radiation, either direct or indirect, for long periods of time, disturbances will occur in the physical balance of the human system. The result will be functional disorders of the nervous system, a vitamin A deficiency, a weakening of the body's defenses, and an aggravation of chronic diseases."*

Research data shows that as light enters the eye, it stimulates vision, but also creates activity in both the pineal gland and the pituitary gland, as well as other areas of the mid-brain region. These two glands control the production of hormones. This means that light exerts a chemical influence in the human body. Dr. Richard Wurtman of the Massachusetts Institute Of Technology has declared implicitly, that light is the most important environmental input after food in controlling bodily functions.

As recently as twenty years ago, textbooks described the pineal gland as a functionless, evolutionary leftover, much like the appendix is described today. Dr. Wurtman, in a ten year study, found that the pineal gland releases a crucial hormone called melatonin. This small reddish grey lump, that sits at the base of the brain, turned out to be not something just "left over."

Melatonin is a crucial hormone that was discovered by Aaron B. Lerner and his team at Yale University Of Medicine, and is regulated by light. Melatonin is responsible for that sleepy feeling we experience in the evening. When light interacts with the retina (eye), brain and special sympathetic nerves, it effectively suppresses melatonin. Dr. Wurtman, who worked as a team with his wife Dr. Judith Wurtman, focused their research on mood fluctuations. This led to the discovery that light not only had a direct effect on the cycles or rhythms of the body and was connected to Seasonal Affect Disorder, but was also connected to premenstrual stress (PMS) and carbohydrate craving. As light enters the body through the eyes, it has a direct chemical reaction with tissues, nerves and the endocrine system.

their regular process of cell division. Hence, the body's chemistry is thrown off, and disorder ensues.

The research of Dr. Wurtman, Dr. Meites, Dr. Ott, Dr. Lewy, and many others too numerous to mention leaves no doubt that as light enters the body, it stimulates the whole system, and keeps the body healthy.

I believe that our lack of exposure to natural light daily, as well as our exposure to imbalanced light sources in our environments, creates an imbalance in our body. There is little argument that light is necessary for good vision and perception. We all interact with these facts daily. From a visual standpoint, indoor lighting is not as effective as it should be; therefore, it certainly cannot be effective in the area of bio-energy.

Most environments today expose people to imbalanced light sources. Incandescent light sources are almost totally lacking in the blue side of light. Most fluorescent lighting screens out one or another wavelengths. In other words each distinctive light source lacks a balance of wavelength (color) frequencies. This equates to our body being bombarded with an abundance of a particular color or wavelength frequency. Different wavelength frequencies are represented in the different individual light sources we use in our environments. Note the spectral graphs shown in Figure 10 and you will see that most of them are lacking in various wavelength frequencies (color).

Most of us spend ninety-five percent of our lives inside buildings or behind the wheels of our cars. Even when we do treat ourselves to a day of relaxation outdoors, darkly tinted sunglasses cut out not only ultra-violet light but all daylighting. This creates what Dr. Ott refers to as "mal-illumination" of the body. The body is being light starved.

You cannot consider light passing through window glass as exposure to full spectrum daylighting. Light is made up of vibrating wavelength frequencies. Glass, although solid to the touch, is made up of vibrating particles. This crystalline substance transmits light, but neutralizes out some of the wavelength frequencies needed. When a wavelength frequency of light vibrates at the same indentifying vibration as the invisible atoms in glass, they will cancel out each other. This neutralization of frequencies is the reason you cannot tan through glass. It is also one of the reasons of why your body does not get the needed nutrients vital for a healthful, energetic life. Dr. Robert Neer, a professor at Harvard Medical School, tested the use of full spectrum lighting indoors on elderly people. The full spectrum lighting dramatically improved their calcium levels.

Normally, cells that make up the body reproduce themselves in an orderly manner so that worn out tissues are replaced, injuries are repaired and then the growth in the body proceeds. Occasionally certain cells undergo an abnormal change and begin a process of uncontrolled growth and spread. One cell will then divide into two, those re-divide into four, etc. These cells may grow into masses of tissue called tumors. Some will be benign, others malignant (cancerous).

Dr. Zane R. Kime, in his book *Sunlight* says, "*Studies have shown that exposure to ultraviolet light or natural sunlight (but not enough to redden the skin) increases the number of white blood cells in the human blood. Amazingly, the white blood cell that increases the most is the lymphocyte. It is the lymphocyte that plays the leading role in defending the body against an invasion of germs. Because lymphocytes increase in number after being exposed to sunlight, their products of defense, antibodies (mostly gamma globulins), also increase in the blood. This increase of lymphocytes and gamma globulins, greatly enhances an individual's ability to fight infection. In fact, it has been shown that experimental animals, which have increased lymphocyte counts, have a high degree of immunity to cancer, as well as tuberculosis.*"

These facts prove that the eye performs a dual function. The first is the role vision. The eye's second job is as a major contributor to human health.

Dr. Ott proved that a person will show a weakening of the deltoid muscle when tested under fluorescent lighting. Under natural daylighting, or under full spectrum lighting, normal muscle strength is restored. I've tested some of Dr. Ott's theories on myself, regarding tinted lenses. I wore eyeglass lenses tinted with a yellow, blue, and rose tint. I wore each individual tint for one month. In my normal daily routine, wearing either the pink or yellow lenses, I felt more tense and irritable. When I wore the blue tinted lenses I was so tired by eleven a.m. that I felt it was almost mandatory for me to lie down.

Light passing through tinted lenses is carried down into the glandular system, creating different reactions, depending on the tint of the lens. If your eyes are overly sensitive to light, and you find it necessary to wear tinted lenses, the best color to wear is grey, which will have no effect on your glandular system and its functions.

Direct influence of daylight on the body's biological processes in turn affects perception, performance, energy levels, and general health. I liken the action and effect of light on the human body to a more refined version of photosynthesis in the plant kingdom. A plant will trap light's energy so it can produce food and grow. This process is called photosynthesis. In this procedure, light is trapped and unites with carbon dioxides and water in the presence of chlorophyll to form simple sugars. These sugars elaborate into starch, proteins, organic acids, fats and other products. Most of these compounds are food for the plant, as well as for animals who feed upon them. Scientific fact shows that the growth of all plants is affected by the length of day, and the intensity of light, even more than temperature and moisture.

Photosynthesis traps the energy of light, in the human body light is converted into chemical energy. In years to come, the platitude of the eye being the window of the soul may prove to be more than just an old maxim.

CHAPTER 8

THE EFFECT OF LIGHT ON COLOR

The way we see color on an object or surface is dependent on two factors; the way the object absorbs or reflects light, and the type of light source that lights the object or surface. The illumination of an individual light source will tend to reinforce certain colors. You might say that each type of light source has specific colors connected to the source. When there is a greater percentage of cool colors present in a light source, it will intensify cool colors and neutralize warm colors. Conversely, when the percentage of color becomes warmer, it will intensify warm colors and neutralize cool colors.

Therefore, a light source affects the success of color in rooms, in make-up used on the mobile canvas of the face, or in matching paints or fabrics. Because different colors do radiate from different types of lights, colors should always be matched or chosen under the same type of light source in which they will be seen. Matching color requires a different technique than choosing colors for harmonious reasons.

Light diagrams (see spectral graphs in Figure 10), are known as Spectral (color) Power Distribution Curves, these provide a visual profile of each color represented in each type of light source. In looking at the diagrams you will note the color differences.

Notice the diagram of a north sky, you will see that there is far more blue in this type of exposure. Incandescent light is very warm, therefore it radiates almost all red and yellow. The basic warm white fluorescent, or the cool white fluorescent, are also imbalanced in the distribution of their color waves. A warm white fluorescent has an abundance of red, yellow and green. The deluxe cool white or standard cool white, which are the light sources most commonly used in public spaces, have an equal distribution of the blue and green, but as you can see on the spectral graphs, they are short in the area of red.

How many times have you tried to find your car in an outdoor parking lot that was lit with low pressure sodium (LPS)? The color rendered by the light source, made it almost impossible to identify your car, based on the fact that the color wavelength frequencies distorted the color. The perception of any color can vary, depending on the type of light source under which the color is seen. The truest color can be seen under daylight. The artificial light source that gives the best indoor color rendition is a quartz halogen light.

The actual duty of a light source, besides giving good vision and acuity, is to represent the color that is seen in it, in as true a manner as possible. The distortion of color waves, radiating from any light source, will change the surface color one sees. At times, as in the instance of quartz halogen, the change will be minimal. At other times, the color can look entirely different, depending on the source of artificial light used. The make-up one uses on the face will change when seen under fluorescent light, daylight, candlelight, or lamp light.

Natural light tends to effect changes in colors, since light changes at different times of the day and year. So as light streams through your window in January, the colors may look slightly different than they will in June. From dawn to dusk, or season to season, natural light changes, which in turn changes color. In the early morning hours, before the sun comes up, light gives off a cool or gray cast. The afternoon sun is intense white light, which incorporates all colors in it, and therefore will tend to intensify color. As the sun sets in the west, colors will take on the rosy type of glow coming from a sunset.

Furniture, fabrics, accessory colors, carpets, and wall color will all look much brighter in direct natural daylight than they will under artificial light. When there is low light, or a lack of light, colors tend to appear darker. It is important when lighting your environment to look at significant questions:

1. How attractive will the color scheme appear in the light in which it will be used? ie., food, make-up, clothes, merchandise, furnishings.

2. How true to color (vivid) will the colors appear under the light source that is being used?

3. How bright or visible will objects appear?

How many times have you selected make-up that looked perfect in a store, only to apply it the following day under the fluorescent lighting in your bathroom, not understanding why your complexion didn't glow, as it did in the store? How many times have you selected a fabric outside your home, matching it perfectly to another swatch of fabric under store lighting, only to find that when the fabric was in place in your room, it looked entirely different? Because of the different lighting in which it was seen, your color changed.

Light and color can be used to emphasize product color, such as sizzling pink light on mauve merchandise, or pools of red light may be used to accent or feature a setting. Ask ten different designers for a color decision in a room, and you may get ten different solutions. Harmonious color selection from a design standpoint is subjective. There are various psychological responses to color which can be either reflexive, cultural, or social. However, when it comes to light and its effect on color, answers to questions are based on hard scientific data.

CHAPTER 9

THE PROFOUND PHYSICAL EFFECTS OF LIGHT

Research studies have shown that the wrong kind of light can lead to physical illness, emotional problems, high cholesterol levels, and tumors. An incorrect lighting situation can create activity in hyperactive children and seizures in epileptics. The significance of the effect of light on the body has been an on-going study for Dr. John Ott and other researchers. Dr. Ott has been the forerunner in this field and continues to carry the torch, hoping to make inroads into medicine and science. Ott's experiments with the CH3 strain of mice, used by researchers because they are susceptible to tumors, showed dramatic responses to colored fluorescent light. Under pink fluorescent light, their tails became spotted and covered with severe lesions, and their tails eventually dropped off. Although their hearts were not exposed directly to the fluorescent light, large dark areas appeared on the hearts of the mice. Under plant "grow lights," which are primarily in the red/blue range, they lost most of their fur, and their tails became scaly.

Ott became convinced from these studies that there were great implications in these findings. Taking his studies a step further, he was able to link a leukemia cluster to children in classrooms where warm-white fluorescents were used in conjunction with closed curtains.

Ott speaks across the country to universities, doctors, designers, and other major groups. He is often called in as a consultant, and in a letter to Ott, Syd Thrift, then Director of the Kansas City Royals Baseball Academy, stated, *"The performance of one of our players, who had mysteriously retrogressed for no evident reason, has, since the removal (Ott recommendation) of a psychedelic-type red lighting from his dormitory room, regained his usual good performance."*

In 1970, the Kline Chinchilla Research Foundation in Utica, Illinois, announced the results of a five-year study in which more than 2000 chinchilla ranchers throughout the world participated. The results showed that when ordinary incandescent light was used in breeding rooms, litters would average 60-75% males, and that by breeding the chinchillas in "daylight" bulbs, the ratio of male to female reversed itself, and the litters produced 60-75% females. Study after study shows that light has a profound biologic effect on animals, and people. In 1969, Wurtman and Weisel demonstrated that the gonads and hearts of rats who were maintained under cool white fluorescent light from birth grew more slowly.

In 1974, two study groups were looked at by Maas, Jayson and Kleiber at Cornell University. They found that students studying under full spectrum lighting showed significant increases in visual acuity (perception), and were much less fatigued than those studying under conventional cool-white fluorescents. Dr. H. Richard Blackwell, a professor emeritus at Ohio State University, found that most indoor lighting diminishes visual effectiveness, in turn impairing working ability. In his studies, he found that light that derives near-equal energy from all bands of the visible spectrum could increase worker productivity by 11.7%.

In 1984, the First International Conference On The Medical and Biological Effects of Light was held at the New York Academy Of Science, based on the fact that the effects of light on the body's biological processes were being affected and needed to be investigated.

Most of us take light for granted. We walk into a room, flick a switch, and voilá, there is light. We don't give it a second thought.

11:00 a.m. your body begins to sag in feeling. A… to lunch and you are refreshed by full spectrum dayligh… lunch. You return to the office feeling good, but by 3:00 p.m. … are fighting sleepiness once again, all created by an imbalance in light wave frequencies, discharges, and an invisible flicker.

In 1975, Bushnell Optical Co., a division of Bausch and Lomb, Inc., in Pasadena, California, changed all of their lighting to full spectrum lighting. Guy E. Bowe, Bushnell's traffic-facilities manager, said at that time, *"We've found an increase in efficiency, absenteeism has dropped, and in the course of a working day, there is less fatigue. People seem to remain as alert and efficient in the afternoon as they are when they come in the morning."*

Lighting becomes even more important when you realize that the greatest part of a work day is spent doing visually-oriented work. Eighty-five percent of all information one receives is received through the eyes. Vision, as well as biological aspects of the body, should be maximized by the quality of light in a lighted environment for optimal function. What is the meaning of "the quality of light?" It has come to mean satisfaction, performance, productivity, and comfort in an environment. It is as important to our health and sense of well being as the food we eat.

In the 1950's, a researcher by the name of Robert Gerard showed that red light caused more tension, excitement and hostility in subjects than blue lights. Red light was found to stimulate the autonomic nervous system. This means that blood pressure, respiration rate, electrical skin conductance, and muscle tone were affected. It also showed that red light made people feel warmer, just as cool light causes people to feel cooler.

Schools, who are educating the future minds of our country, need to be more aware of the role that light is playing in education. Using full spectrum lighting in schools showed that such lighting reduced the incidence of colds and viral infections.

Dr. Ott conducted a six month study at the Gocio Elementary School. He lead shielded and installed full spectrum lamps in

... in the school. He then mounted
...s on brackets from the ceiling. The
...on/off at different times when the class
...t the students knowing it, during various
...(time lapse photography). The students
...fferent under the lead shielded fluorescent
...had previously. One child who was hyperactive
and ...tory in class became interested in class work and
moved ... front of the class. The other children appeared
calmer and ...ar more interested in what was going on in class.

Being alert and ready are the priorities in police work. In 1987, the Police Department in the city of Dayton, Ohio, was able to install new fluorescent fixtures manufactured under Dr. Ott's patent (Ott Lites) in a new building that was a dispatch center. After two years, they reported no eye strain or headaches from glare-free video terminals that the police work at daily. Although an intensive 24 hour operation center, there were no complaints.

For offices, the bottom line is raised productivity and quality of work. In 1966, Zamkova, in a study using full spectrum lighting, reported that individuals had lower fatigueability and significantly improved working capacity and performance.

Ott Lites are full spectrum fluorescent fixtures that are lead-shielded to prevent radiation from discharges. They have been manufactured with modified fluorescent tubes. These have spectral characteristics that most closely simulate full spectrum daylighting, and provide the favorable effects needed for alertness, performance, and productivity.

Although man has achieved wonders in this progressive society of ours, the quality of light in our school systems and work spaces presently do not lend themselves to optimal conditions for our body's sake.

Ancient man symbolized nature, the elements, gods, good and evil, the four points of the compass, and religion through the use of color. Color symbolized night and day, the earth, sun, fire, and the stars. The Greeks believed that all matter was composed of four elements. These were symbolized with color. Earth was green, air was yellow, water was blue, and fire was red.

Purple became the Royal color of Rome, the son of a prominent parent wore a toga with a purple border. This was because purple was the most difficult color to duplicate. Thousands of snails lost their lives so that a minute amount of purple dye could be reproduced for material that was worn by reigning monarches and those of prominence.

It was in ancient Egypt where cosmetics were first used, not used just to enhance beauty, but to reinforce racial differences. The Egyptians believed that God the creator painted the skins of different races in different colors. The ancient Egyptians considered themselves to be of the red race and thought their skin color was the reflection of God's creation. They used Henna to stain their bodies red, enforcing God's message, and to show pride in their skin color. In all of their pottery and art, you will see this color used as part of a color statement made by an ancient culture.

Today here in America, color still symbolizes a message of racial pride. The message is related in the tone of skin color. The Black person whose skin is darker, is accepted as more racially pure, and therefore more accepted by other Blacks. There are twelve different skin tones in the Black culture. These skin tones seem to differentiate each person according to color and social acceptance. For instance, a Black person of mixed color, with light skin tones, is often shunned socially by his own race. In a tribe known as the Ndembus, women with very black skin make the best mistresses, but they are not taken as wives.

Color does not play a big role in Christianity today; however, in early Christianity, the trinity was associated with the colors of blue, yellow and red. White indicated purity and joy, red related to charity, blue represented truth and divinity, green was the color of hope and life everlasting, purple stood for sorrow and suffering, and black was the color of death. There are numerous references to brilliantly colored silks, a mention of a red garnet at the bow of Noah's ark, which was to light and guide his way. Additionally, the rainbow, Biblically, is considered the symbol of man's unity with God. Blue in modern times still represents peace, truth, and divinity. You will find that it is used as the color of Mary's robes whenever she is depicted, and it is especially shown on present day religious greeting cards.

Ancient man never thought of changing symbols or color, any more than they thought about changing social structure or religion. In primitive times, all were related. Color was not used freely because it met a function. Today, color symbolically is seen differently, depending upon the individual. Take red for instance; red may be seen in association with fire, the sun, heat, sensuousness, devilishness, hell, ego, or indebtedness.

No matter the individual, one connects colors constantly to left-over ancient perceptions, superstitions, and thinking. A black cat that runs in front of you is bad luck, white represents purity, and green is the color of freshness and new growth. Traffic lights tell us to stop, wait, and go, sending a silent, but clear statement of the effects of symbolism, and telling us that color has a language all its own.

PERSONALITY TRAITS AND COLOR PSYCHOLOGY

From the standpoint of psychology, there is great value in what color reveals about the character and personality traits of an individual. Ever since Dr. Max Luscher, a Swiss professor of psychology, refined his Color Test in 1947, his information has been used widely. Using the Luscher Color Test correctly, one can assemble a wealth of accurate information about the conscious, as well as unconscious, psychological make-up of a person. This information may even be translated for groups of people, or buying segments in order to sell products.

Your eyes and your brain help you to experience light and color. Your vision accumulates observations that become data. This information about people, the environment, and objects then helps you to formulate ideas which are transported to the brain. The upper two-thirds of your brain is where your ideas take shape and become perceptions. Without the help of the eye and brain feedback, we would not be able to see color or attach perceptions to it.

The area of the world you live in, both geographically as well as regionally, your age, and the status of your socio-economics are the determining factors where the perception of color is concerned. Socio-economics deals with more than just money. It deals with the level of education you have achieved, life values, and life style.

People who fall into the bracket of higher socio-economics will tend to respond to more complex mixtures and sophisticated colors. These are colors such as a blue shade of grey, deep blue-green, deep burgundy, taupe, etc. People from a lower socio-economic status will tend to respond to less complex mixtures and colors that are generally brighter and stronger. In the instance of orange, whether it is a strong orange or a muted orange, this color seems to appeal to those in a lower socio-economic class.

Attitudes toward color will vary from person to person. You can be sure that as a person's economics, life style, and life values change, his favorite color will change along with them. As traumatic happenings affect your life both physically and emotionally, you will find that your choice of a favorite color will begin to change afterwards.

Rorschach found that a cheerful person responded well to color, whereas depressed people more often reacted to shape or form. Color dominance indicated an openness to any external type of stimuli. Rorschach felt that such people were sensitive, easily influenced, unstable, disorganized, and given to emotional outbreaks. A preference for shape or form in patients, he said, showed an introverted disposition, strong control over impulses, and an unemotional attitude.

Scientific investigation has shown that adults associate mood and color. Both active and passive values are related to colors. Deborah Sharpe, in her book, The Psychology Of Color And Design, writes about a study using college students as an index in associating mood to color. She said that the study showed that the warm colors of reds, yellows, and oranges were connected to action, aggression, excitement, and stimulation, while the cool colors of blue and green were associated with the feelings of being calm, secure and at peace. The colors of brown, black and grey were associated with melancholy, depression, and sadness. Yellow related to cheer, happiness, and fun; and purple was associated with dignity, royalty and sadness.

C. W. Grant did a doctoral dissertation on subjects that were neurotic and psychotic. Subjects who tended to be neurotic chose red and yellow as their favorite colors. Subjects with tendencies toward psychosis had an affinity for blue and green.

People tend to see the world in their own terms. Their vision is colored (pardon the pun) by their formative years, their environment, and their experiences in life. No two people come from the same background, or have the same experience; therefore, no two people will have the same attitudes and perceptions. Think back on your own life from your earliest recollection up until today. Take a sheet of paper and draw four columns (see illustration on following page). Chart your age in the first column. Write in your favorite color for that age in the fourth column. In the second and third column, write in events that impacted your life strongly. You will be able to relate both positive and negative events to the effect of a color change. These might be happy events like a marriage, birth of a child, or a promotion to a top CEO job. They could be sad events such as a divorce in the family, death of someone close to you, a life threatening surgery, near-death experience, loss of a job and extended time off work.

Because life places stresses in our everyday lives, these stressors help to direct our attitudes, perceptions, and color choices. Your instinctive selection of a favorite color at a given time in your life will indicate your conflicts, balance, and emotional disturbances.

There are both positive and negative qualities in all colors. Because people are unique, they will have both positive and negative characteristics to their personalities; these tend to match to the positive and negative characteristics of a color.

The psychological testing that has been done on color is too vast to mention in a book this short. The information in this chapter, on the meaning of different colors, is based on research of professional psychologists. It is a broad overview and not meant to prejudice. Since color psychology is not an exact science, and is based on personal interpretation, the reader should remember that selection of a favorite color is also subject to hidden traits and hidden desires on an individual's part.

Can you love two different colors and be able to define your personality? The answer is yes. You may love a color that is a combination of two colors, such as blue-green, or you may have two separate favorite colors, such as burgundy and grey.

In the case of a color, such as blue-green, your personality traits would be drawn partially from the two colors of blue and green. In the instance where a person favors two distinct colors, the traits of one color will tend to compensate for some of the traits of the second color of choice, subsequently showing up in personality characteristics.

AGE	POSITIVE	EVENTS	NEGATIVE	COLOR

Negative Characteristics

Less friendly than red, you still attract others to you. More controlled, and with more conscience than red, these controls put a damper on activities that red tumbles headlong into.

PINK

Positive Characteristics

Pink is a tint of red. It is pink that is young, innocent and naive. You who prefer pink see life in that aspect. According to Faber Birren, color expert and author, pink tells of doting parents, wealth, social advantage and a sheltered life. Generally speaking, little girls surrounded by pink bedrooms tend to be sweet and more even tempered.

Negative Characteristics

Because of naiveté you want life never to be unseemly, always to be a Utopia, and when it isn't, you can't understand why. Because you are an incurable romantic, you will always look for the perfect romantic relationship and may find happiness with no one. The first time that any intensity of feeling is shown in the relationship, or the light of your life shows a human failing, you will leave to find romantic perfection.

MAGENTA

Magenta is a mixture of red, blue, and yellow. You will have the energy and passion of red. You make a good administrator, with the ability to organize from the quick intellect of yellow. You are a lover of beautiful things. You are loyal, make a good friend, and are a spiritually oriented person.

BLUE

Positive Characteristics

Blue is the all-American favorite color. It is the color of introversion, loyalty, deep bonds, and a calm life. You with a true blue personality seek tranquility in all that you do. Blue is associated with love, trust, devotion, and depth of feeling. You are a person who has great empathy for your fellow man. Give you a job to do, and you get it done. You are responsible and work hard at anything you undertake. You make a wonderful friend. You don't have a wide array of acquaintances, but have a delightful time with a narrow circle of friends. If blue is your choice in a favorite color, you are conservative and have great control over your emotions. Therefore, in love relationships, your control makes you seem surfacely cool. When the choice of blue goes to a deep navy color, besides being a higher socio-economic color choice, it tells of more than just personality traits; it also relates to a person with lasting values.

Negative Characteristics

You tend to be egotistical, and demanding of yourself, as well as others. Although you may not show it, stupidity or slowness in others is irritating to you.

GREEN

Positive Characteristics

Green is a sign of new growth, freshness, and a balance in nature. You who choose green also seek a balance, a balance between the material aspects and the spiritual side of life. You have a need to understand everything that goes on around you. If you are a person who loves green, you are highly determined, patient, attentive to detail, dependable, and serious. You want life to run the way you think it should run. You are social but bound to very traditional thinking. You are outspoken, and because you are a good friend, you have many friends, taking on the burden of each. You are concerned about always doing the right thing, and ever careful of your reputation and social standing. You are loyal and a good mate.

Negative Characteristics

You can be dogmatic, and you have a need to impress people. You like to eat and therefore have a tendency to be overweight. Because an orderly life gives you security, you will resist change and have great difficulty accepting anything new or risky. You have a constant need to be with people, and desire that they reaffirm their affection for you.

BLUE-GREEN

Positive Characteristics

Blue-green comes in many shades. It may be turquoise, aqua, teal, or peacock blue. Whatever name it wears, it is still blue-green. The lover of blue-green is much like the peacock who wears this color proudly; you will always be primping. You are fastidious in your dress and ways. You love the best in fashion, food, wine, and travel. Nothing is too good for you. You are sophisticated, witty, and charming. People are very attracted to these traits. Good at heart, courteous, and helpful are your outstanding traits.

Negative Characteristics

You are the real Narcissus or title holder of "self-love". Because you are, it is hard for those who love you to get very close. You are busy loving yourself. You love to be admired and feel you are entitled to any attention you get. You fuss over anything that is out of place or any detail that has not been taken care of. You are demanding of yourself and demanding of others.

YELLOW

Positive Characteristics

Yellow is the sign of intellect. It is a luminous color that sets it apart from other colors. If you love yellow, you also want to be set apart from others. You have a superior mind, and a sparkling personality that attracts people to you. Your wide circle of acquaintances gives you an outlet for self-expression and expansiveness. You love to learn, and press forward toward the future through what you learn.

Negative Characteristics

You scatter your energies, and most likely leave tasks unfinished because of it. Whenever the realities of life appear too harsh, or your freedom is interrupted, you will whine and complain to those who will listen.

PURPLE

Positive Characteristics

Purple is the color preferred by artists, philosophers and mystics. Introspective, creative seekers of illusive philosophy, deep thought, and the unusual will choose purple. You are the people seeking a magic kingdom or a dream world where the hidden, the occult, or man's reason for being is discovered. You have the ability to charm the feathers off a bird, but don't often expend the

BROWN

Positive Characteristics

Brown is a color that is associated with the earth. You are restrained, steady and reliable. You feel that the controls that you place on yourself will lift you above the average person. People can't seem to meet your high standards, but you look to them for admiration and appreciation as a way of building self-esteem. You obstinately stick to your opinions. This helps you to compensate for the lack of self-esteem you feel. This in turn causes considerable stress. Because you find difficulty in finding those with standards as high as you feel yours are, you feel alone. You are obstinate, sticking to your opinions right or wrong.

Basically you are a gentle, caring, sensible, and sensitive individual who responds well to gentleness and sensitivity from others. You love your freedom, and balk at anyone who would inhibit it in any manner. If you become disappointed, it will lead to indifference on your part. Although you are not highly intellectual, you are able to think things through clearly.

Negative Characteristics

The suppression and restraints you place upon yourself lead to anxiety. Worry and anxiety hide under your cloak of self-reliance. When you feel you are unappreciated, you become angry and depressed. You feel insecure as time goes by because of a lack of allies. Anyone who criticizes you comes under your scorn. Because you are obstinate about your opinions, anyone who disagrees with you is completely wrong. Your closed mind does not allow you to see another person's point of view.

GREY

Positive Characteristics

Grey is a color that does not recede or intrude. You are a person who has experienced both physical and emotional trauma in your life. You have withdrawn from the mainstream to rest and recoup. You feel life has more to offer than what you have experienced in the past; therefore, you pursue your objectives in a self-learning mode. You are ruled by controlled reason. You no longer want to carry banners, wave flags, join marches, or participate in group activities, such as clubs, etc. It is difficult for you to involve yourself with another person to a point of attachment, yet you need to have a sense of belonging. You feel weary and have a need to be secure and protected. Life's ups and downs have led you to seek a calm existence. You are a hard worker, but stay away from competitive struggle.

Negative Characteristics

You want to experience life, but your control and restraint restrict you. You are restless and impatient. Your need to maintain a peaceful life, and a non-committal attitude, make you seem cool on the surface.

BLACK

Positive Characteristics

Black for the most part is perceived as that which is negative, such as bad luck, death, fear, or mystery. Symbolically black refers to the male. You won't find too many people who will claim black as their favorite color. You want to be unhampered and without restriction. You want to control your own destiny without feed back from those around you. This creates an ambivalence for you, since you also want to overcome a feeling of being separated from others. When it comes to your perception of life, you carry a constant feeling of "Is that all there is?" You are intense, committing yourself in every way to any objective you undertake. Any type of situation that is disagreeable will give you the feeling of being trapped. You are prone to becoming enthusiastic. Witty, made wise by the school of hard knocks, and very private about your personal life are descriptive terms that explain you.

Negative Characteristics

Intensity in all that you do creates stress, eventually leading to a breaking down of your body's defenses. Unfulfilled hopes and dreams have left you feeling emotionally dissatisfied, which creates a restlessness in you. When you become involved in a relationship, you constantly demand to know where you stand. This constant demand, if not watched will cause the other person to eventually leave.

WHITE

There isn't much psychological information connected to white. Primarily, one finds symbolic associations where this color is concerned. Symbolically white is connected to light, and therefore, brightness, and refers to the female. White is also likened to a mystical life, or close union with God. Plutarch associated white with cleanliness and purity of a religious kind. In ancient times, white was seen in conjunction with Olympians (strength); for poets it was a mark of divinity and beauty. From the stand point of mysticism, the color was used as magical protection against evil.

If we were to follow the tradition of linking the meaning of symbolism to color psychology, then white would reveal only positive characteristics. We would find that if you chose white as your favorite color you would be pure of heart, strong in your convictions, and highly spiritual in nature. Since no solid psychological research can be found on the color white, we cannot assume that what the author has written is solid information; it is only speculation on her part.

COLOR PERCEPTION

The professional color consultant, working in areas of design, marketing, trends, and education is aware of the significance that color plays in all fields of endeavor. It is in the profession of color marketing (sales of products through color) that color on packaging projects a message to gain disposable dollars. Your dollars are spent to buy products. Whether in the manufacture of automobiles, food products, cosmetics, fashion etc., each product uses color to portray an image that the retail public will buy. Success is based on the correct color being used to create the right image for a particular social group. Colors can be vibrant, aggressive, cool, warm, violent, exciting, tranquil, or sophisticated. Certain colors are used to market to men, while others speak to women, or teenagers.

The kind of color you would use to market dream vacations is far different from those that would be used to instill confidence in a banking institution. The color message that speaks to your perception is received by you unconsciously. Color not only must be correctly chosen for packaging, but must be well balanced and decorative. Your acceptance or rejection of a product, will be heard loud and clear in sales volume at the cash register. It is no accident that the myriad of colors you see on store shelves were put on packages to attract your dollars and make you buy. Color incorrectly used on a commodity will not have impact on a store shelf. If a product does not stand out, the lost impact will mean lost dollars.

Modern packaging must be appropriate to the product. Color selection used on packaging is based on class, group age, culture, and wealth. Over the years, manufacturers will tend to change the

Market research showed that a detergent that was test marketed in the color blue and yellow was believed to be of top strength. Because of that testing, blue and yellow was moved onto detergent packaging. Other colors you will see used on detergent packaging are red-orange and black. Black relates a message of stability and strength. Red-orange is easily seen on a shelf from a distance, and reinforces the message of black by contrast.

Some food packaging needs to say I'm cool and refreshing, while other colors define an organic health message. If you scan your dairy department, you will notice that there is no green shown on the packaging. That is because green relays a message of decay. Blue and white are always seen in a dairy case, because they suggest hygiene and coolness. Green is used universally in the packaging of frozen vegetables because its message is one of freshness, and green is the symbol for fertility.

Children love primary colors, so drinks for children are color coded according to the flavor that they like. Association of color is so strong for children and adults alike, that even when drinks were lime flavored, but colored orange, both believed they were drinking an orange drink. Small chocolate candies that are manufactured in a number of different color coatings are all flavored the same, yet both adults and children eat their favorite colors first.

It is no accident that bakery goods are put in pink boxes. Bakery goodies in a pink box are perceived as being sweeter. For ice cream, blue, pink, yellow, and soft browns are at the top of the charts.

The one thing we know for sure, no matter the product, loud or soft, warm or cool, every color has its say.

CHAPTER 11

THE FUTURE

In the future, light and color will play a more major role in schools, in the workplace, at home and in the far reaches of outerspace.

With present medical treatment, drugs and hospital stays becoming exorbitant, we will have to look at applications that increase well-being. Color and light can be one of the key drivers in preventive medicine.

In the past years, average school grade levels have fallen drastically. Having taught on a high school level, I can tell you that a teen-age attention span is short, and retention of material is difficult at best. The time has come for school administrators to look at the use of full spectrum lighting in their facilities. Scientific research by well known professionals, on the effect that light has on the body, is raising consciousness levels and should be giving a new awareness to the importance of light in the indoor environment of a school.

Many connections have been drawn between light and metabolism, sugar balance in the body, sexual function, and the effect of full spectrum lighting on blood counts. Because of this, education needs to take a harder look at the effects of standard fluorescent lighting that is present in all educational facilities.

In the October 26, 1987 issue of Time Magazine there was a large article on "A Day In The Life Of The Soviet Union." It was interesting that on one of the pages there was a picture of Stavropol children being exposed to light indoors to prevent vitamin D deficiency.

There is no reason that quality, full spectrum lighting cannot be used in schools, which in turn would have a direct effect on the productivity of both student and teacher alike.

"A room lighted with fluorescent lighting is less than one-tenth as bright as the area under a tree on a bright sunny day," says Dr. Robert E. Jenkins, Director of the Clymer Health Clinic, in Pennsylvania. We are now at a stage that we should be concerned about the missing ingredients of light that neither incandescent or fluorescent light give us.

Competition in health care presently is at its highest peak. Hospitals are now competing for patients. This will become an even more looming problem for health care facilities within the next five years. Hospitals in the future will have guests; doctors will have patients. Eventually hospitals will become critical care centers only. Today, a priority for any health care environment is a surrounding that encourages a *"Get Well"* message in order to attract paying patients.

Integrating light and color allows for an element of change in sterile hospital surroundings. Colored sleeves that fit over fluorescent lamps is an inexpensive way to bring color into a hospital room. Rapid convalescent statements can be made through color on surface finishes, in window treatments, and in unsterile looking colored curtains that at times surround the bed.

Light and color have a bright future for hospital and office corridors. We'll see glass sandblasted and carved, lit with changing colors by fibre optics. Large engraved glass art panels will also be used in restaurants and residences.

Louis Kahn, the great architect, said, *"All architecture begins with light."* Kahn believed that although light did not make the architecture, it was a servant to it. More and more architects are folding Louis Kahn's precepts on light into their architecture, along with color. In the future we'll see more commercial buildings outlined in light and color through neon and fibre optics. Architecture is forms and planes. A building is an architect's art composition. Light performs the task of assisting by reinforcing the architect's composition.

For the last ten years I've said that the wave of the future in both residential and commercial buildings will be to paint with light and color. Light and color will be projected across walls to create illusion, and holograms in light boxes will bring us the third dimension. Holography is a unique photographic process that records an image onto a plate. When a point light source meets on that plate, a three dimensional effect is achieved. This allows for the possibility of having great works of art or sculpture in one's own environment.

From Earth to the last great frontier of outer space, light and color will be present to play an important role. Futurists envision outer space housing as a progression from an eight person operational center to towns of 100, or colonies of 10,000. But in the beginning, a small number of modules that hold about seven people will connect like a triangle, or square. When outer space is in its formative stage of being colonized, human beings will live in tight quarters in a harsh environment. The brave souls who take up residence in this new world will live and work in a sustained monotonous surrounding. In this type of habitat, light levels which have a direct affect on behavior will be most important. In 1974, Sanders, et al, did a study and found that in humans, if light levels were increased, sound or noise levels were increased.

Light levels are only one consideration. In 1978, Rogers found that decor in a module had little value to crews involved in meaningful work. When crews were composed of both males and females, even though the crews had meaningful work to perform, decor became important. In 1973, Berry reported that the sameness of color within the interior of the module was disturbing to astronauts. This tells us that light and color, and their effects will play an important role.

One of the roles color will play is in coding for "cognitive image." This is a psychological term. Here on Earth its meaning refers to how people find their way from one point to another in a building; it also helps persons determine how they feel about the space they are occupying. In other words, how persons might find their way in an emergency is important when it comes to a means of exit. In the 360 degree directionless world inside a module, color and light could code, providing a speedy way for astronauts to make quick visual judgments and provide them with needed spatial referencing.

When a group of people is confined to a small area for 90-120 days, the psychology of living changes. Personal space (imperative territory) becomes a priority. The face of locker doors, behind which a small compartment holds personal belongings, could be finished in different colors. Not only would it create a pleasant wall of color coded squares, but it would act as personal identification, for an individual astronauts' belongings.

A loss of calcium in bone and muscle mass is one of the reasons that astronauts spend only a maximum of three months in outerspace. All but ten percent of the calcium is naturally replaced upon the astronauts' return to Earth. In the overall body picture, ten percent is very meaningful. Sleep areas could be fitted with small light weight color lamps (color therapy), to assist with the prevention of calcium loss. This of course would need to be tested in controlled studies.

The future is bright and promising for light and color; how one sees it, of course, is in the eye of the beholder.

INDEX

Color Structure and Design Richard G. Ellinger
Van Nostrand Reinhold Company New York 1980

Color Symbolism Benz, Portmann, Izutsu, et al.
Spring Publications Dallas, Texas 1977

Color Therapy R.B. Amber
Basusree Press Calcutta, India 1964

Color Your World Frank Don
Destiny Books New York 1977

Colors and Numbers Louise L. Hay
Hay House Santa Monica, California 1987

Creative Color Faber Birren
Van Nostrand Reinhold Company New York 1961

Electro Therapy and the Elements of Light Therapy
Kovac, RichM.D. Lea and Febiger Philadelphia 1932

Eye For Color Bernat Klein
Pellans and Wilson Limited Scotland 1965

Healing Through Color Theo Gimbel
C.W. Daniel Company England 1980

Health, Youth and Beauty Through Color Breathing
Linda Clark and Yvonne Martine
Celestial Arts California 1976

Is Your Volkswagen a Sex Symbol Jean Rosenbaum, M.D.
Hawthorne Books, Inc. New York 1972

Joy's Way W. Brugh Joy, M.D.
J.P. Tarcher, Inc. California 1979

Light Richard Morris
The Bob Merrill Company New York 1979

Light, Color and Environment Faber Birren
Van Nostrand Reinhold Company New York 1982

Light, Radiation and You John Ott
The Devin Adair Company Connecticut 1982

Man's Guide to Science Isaac Asimov
Basic Books, Inc. New York 1960

Myths of the Sun William Tyler Odcott
Capricorn Books New York 1967

Photosynthesis Isaac Asimov
Basic Books, Inc. New York 1968

Rainbow of Liberated Energy Ngakpa Chogyou
Element Books Limited Longmead, Shafesbury, Dorset 1981

Rainbow of Light L. Harris
Joysun publications Hawaii 1980

Report On Radionics Edward W. Russell
The Priory Gate England 1979

Seven Mansions of Color Alex Jones
De Vorse and Company Marina Del Rey, California 1982

Spectro-Chrome Metry Encyclopedia Dinsha P. Ghadiali
Three Volumes, Spectro Chrome Institute New Jersey 1939

Sunlight Zane R. Kime, M.D.
World Health Publications California 1980

The Ancient Art of Color Therapy Linda Clark
Simon and Schuster New York 1975

The Art of Science and Color Hal Hellman
McGraw-Hill New York 1967

The Curative Value of Light Edgar Moyer
D. Appleton and Company New York London 932

The Elements of Color Johannes Itten
Van Nostrand Reinhold Company New York 1970

The Healing Power of Color Betty Wood
Destiny Books New York 1984

The Human Aura Walter J. Kilner
Citadel Press New Jersey 1965

The Luscher Color Test Ian Scott, Dr. Max Luscher
Washington Square Press New York 1969

The Magic of Color Thomas H. Wolf
Odyssey Press New York 1964

Theory and Practice of Color Frans Gerritsen
Van Nostrand Reinhold Company New York 1975

The Physics and Chemistry of Color Kurt Nassau
John Wiley and Sons New York 1983

The Principles of Light and Color Faber Birren
Edwin S. Babbitt (Deceased) Citadel Press New Jersey 1967

The Promise of Kirlian Photography Mikol Davis Earle Lane
Harper and Row Colophon Books New York 1978

The Science of Color Optical Society of America
Thomas Y. Crowell Company 1953

The Spectro Chrome System Darius Dinshah
Dinshah Health Society New Jersey 1979

The Universe of Light Sir William Bragg
Dover Publications New York 1959

The World of Light John Stewart Colles
Horizon Press New York 1960

Ultraviolet Radiation and Actinotherapy
Eleanor H. Russel, M.D., B.S., W. Kerr Russell, M.D., B.S.
William Wood and Company New York 1928

Vibrations Virginia MacIvor, Sandra LaForest
Samuel Weiser, Inc. New York 1979

You Are A Rainbow Norah Hills
University of Trees Press Boulder Creek, Colorado 1979

CLINICAL PAPERS

Seasonal Affect Disorder A Description Of The Syndrome
Rosenthal, Sach, Gillia, Levy, Goodwise, Davenport, Newsome, Wehr
Archives of General Psychiatry
Clinical Paper March 1983

Seasonal Cycling In A Bipolar Patient Rosenthal, Levy, Wehr, Kera, Goodwin Psychiatry Research pp. 8, 25—31 1983

The Therapeutic Value Of Light And Color Kate W. Baldwin, M.D. F.A.C.S. Former Senior Surgeon Woman's Hospital Philadelphia, PA
October l2, 1926 Clinical Meeting - Eye, Ear, Nose and Throat Medical Society of Pennsylvania

COURT TRANSCRIPTS

Kate W. Baldwin, M.D., F.A.C.S.
Excerpts from Transcripts of Litigation Trial against Dinshah P. Ghadiali
Buffalo, New York 1931
Senior Surgeon Woman's Hospital
Philadelphia, PA

MAGAZINE ARTICLES

Changing Hospital Environments Dorothy Kamm
Lighting Design and Application November 1985

Light In The Laboratory Sarah Shaukman
Lighting Design and Application May 1986

TREATISE

A Course In Specific Light Therapy Carl Loeb, M.D.
Actino Laboratories Chicago 1939

RESOURCES

Color Marketing Group of the United States
4001 North 9th Street, Suite 102
Arlington, VA 22203
(703) 528-7666

Ott Light Systems, Inc.
306 East Cota Street
Santa Barbara, CA 93103
(805) 564-3467

Address all inquiries to

Barbara Colby
c/o Chroma Productions
~~501 W. Glenoaks Blvd., Suite 30~~
~~Glendale, CA 91202~~
(818) 242-8788

CHROMA PRODUCTIONS ®
BARBARA COLBY
101 E. STOCKER ST.
GLENDALE, CA 91207
(818) 242-8788

NOTES

FIGURE 1

FIGURE 2

FIGURE 3

FIGURE 4

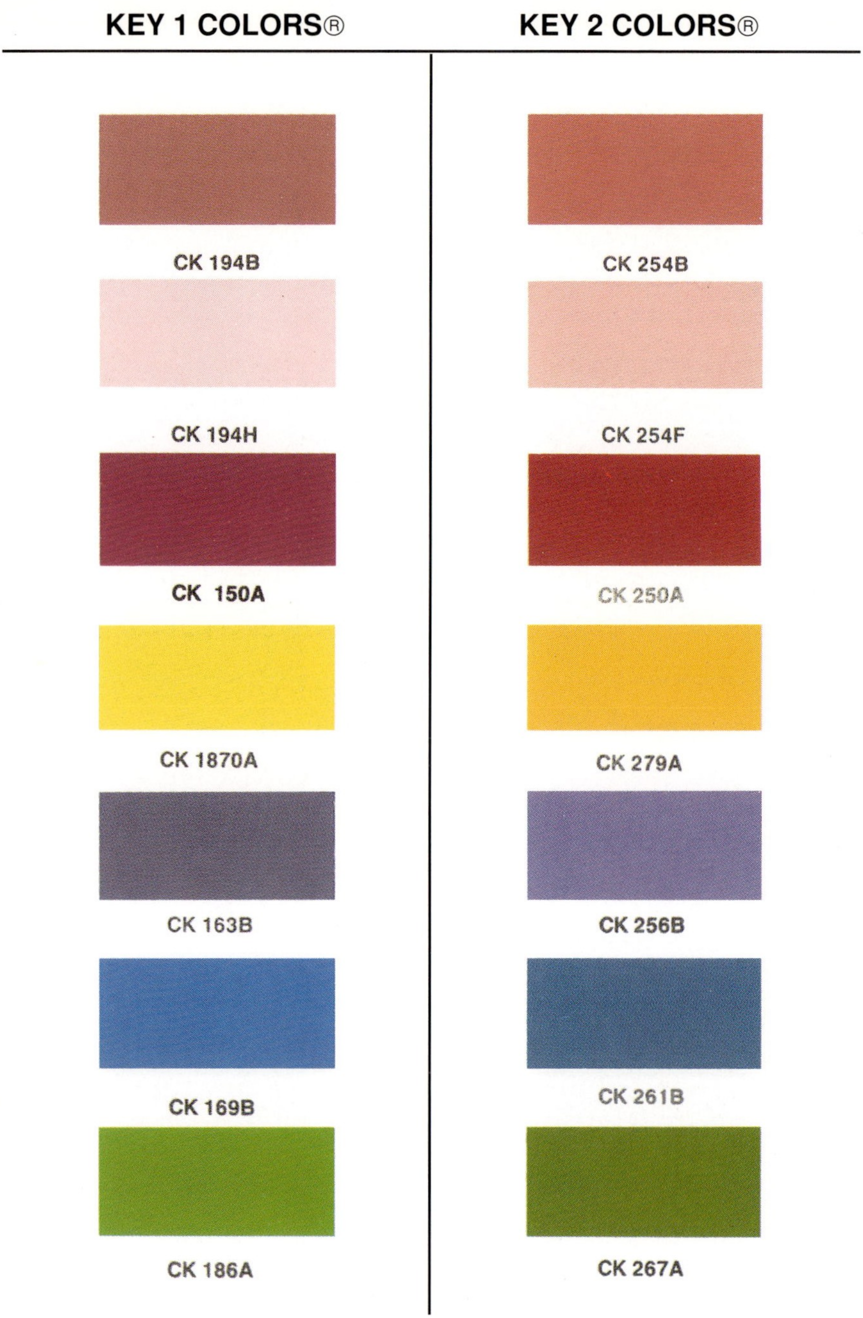

KEY 1 COLORS®
KEY 2 COLORS®
CK 194B
CK 254B
CK 194H
CK 254F
CK 150A
CK 250A
CK 1870A
CK 279A
CK 163B
CK 256B
CK 169B
CK 261B
CK 186A
CK 267A

FIGURE 8

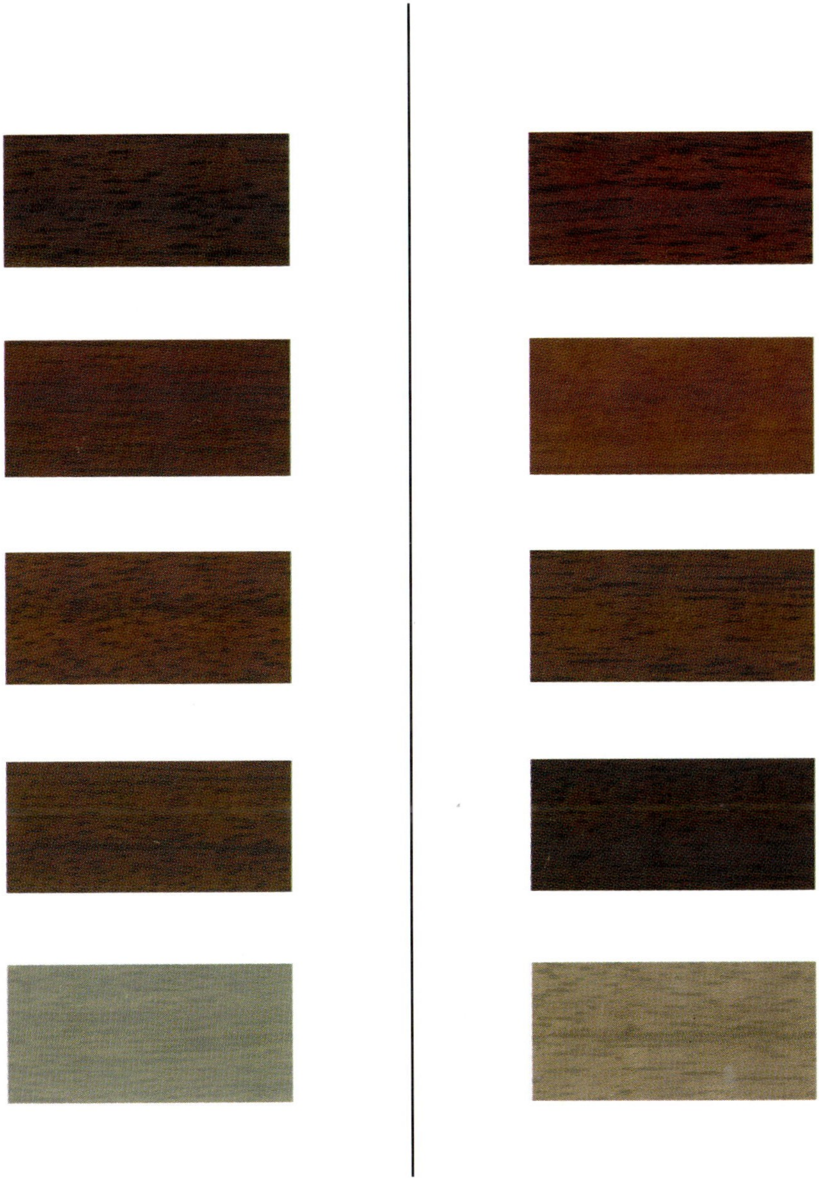

WOOD TONES

KEY 1® KEY 2®

FIGURE 9

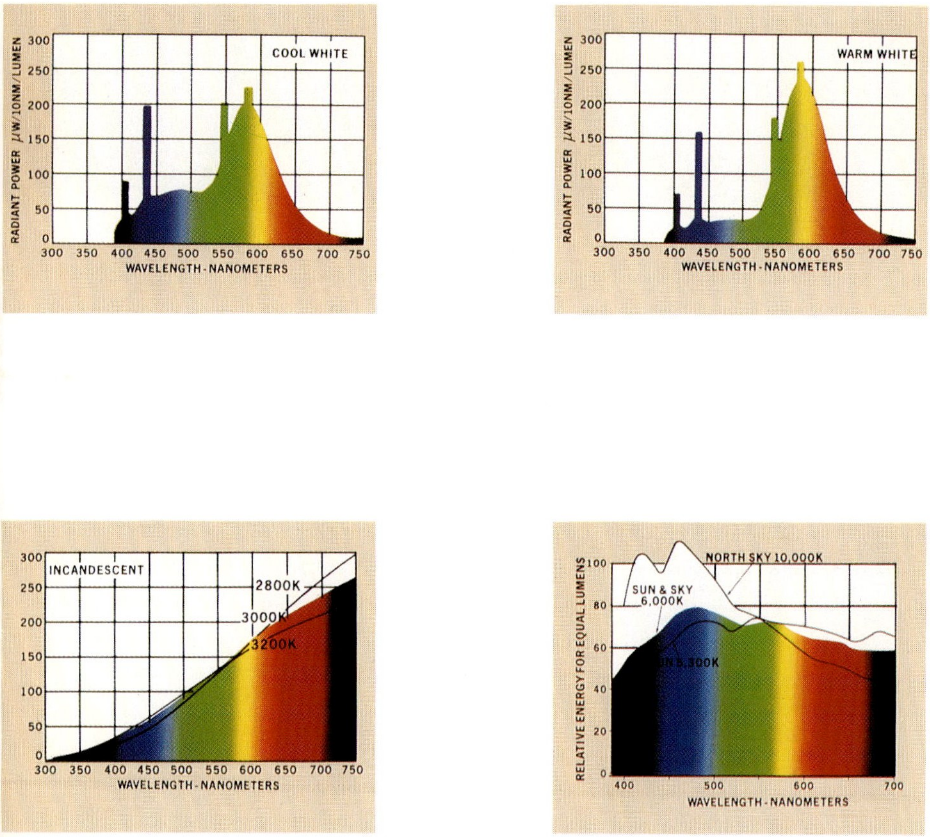

Spectral Graphs

GENERAL ELECTRIC COMPANY, NELA PARK, CLEVELAND, OH

FIGURE 10

TRAVELING WEST (To Earlier Time Zone)

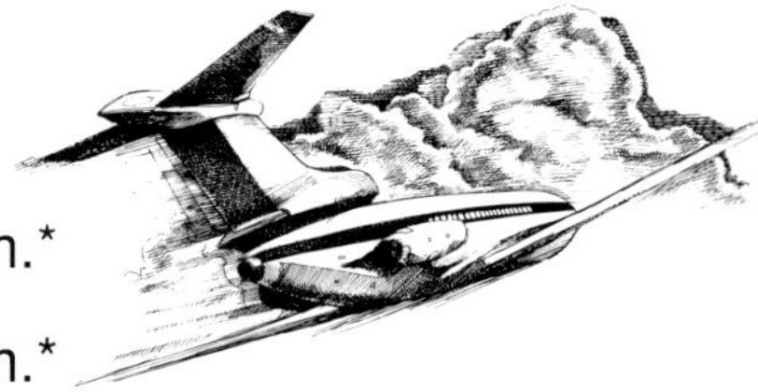

Time Difference	Get Outdoor Light From
2 hrs.	4 p.m. - 6 p.m.*
4 hrs.	2 p.m. - 6 p.m.*
6 hrs. plus	12 p.m. - 6 p.m.*

TRAVELING EAST (To Later Time Zone)

Time Difference	Stay Indoors From	Get Outdoor Light From
2 hrs.		6 a.m.* - 8 a.m.
4 hrs.		6 a.m. - 10 a.m.
6 hrs.		6 a.m. - 12 p.m.
8 hrs.	6 a.m. - 8 a.m.	8 a.m. - 2 p.m.
10 hrs.	6 a.m. - 10 a.m.	10 a.m. - 4 p.m.

**Table assumes sun will rise at 6 a.m. and set at 6 p.m. at new time zone. If it doesn't, readjust table based on actual times.*

Jet Lag Chart

FIGURE 11

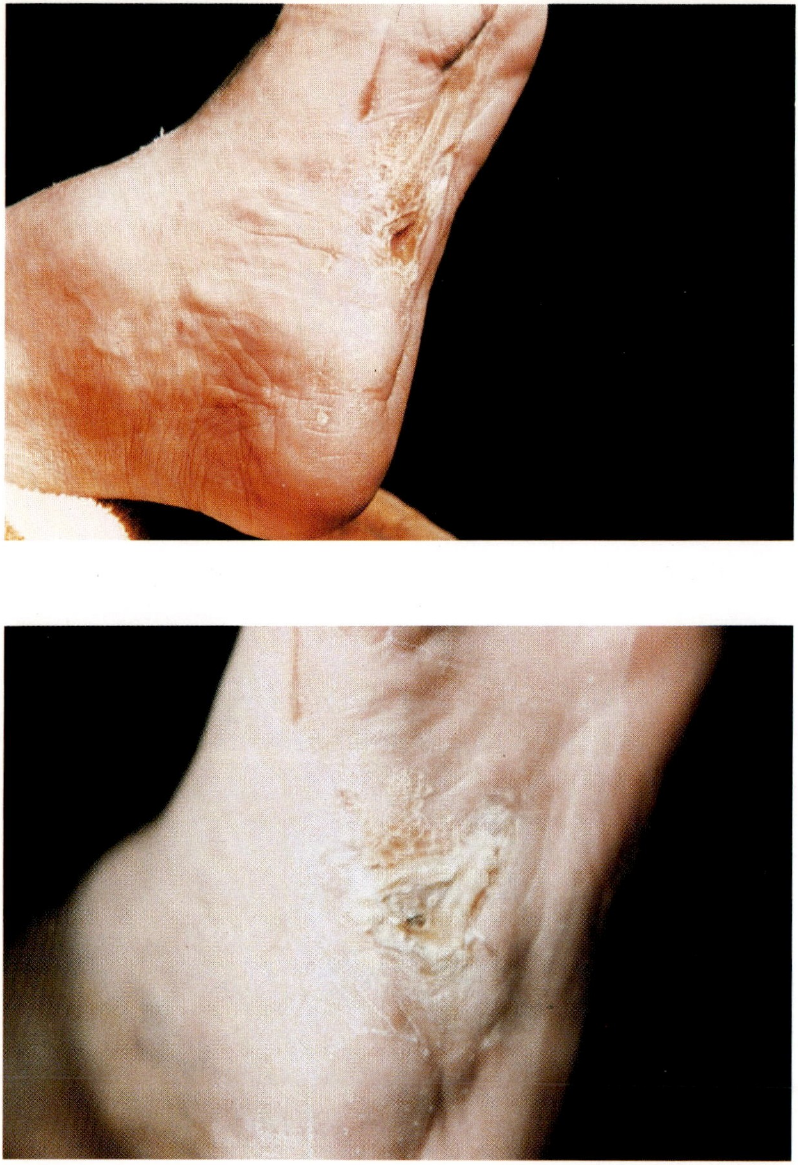

FIGURE 12

Decubitus ulcer shown on foot on top photo taken on January 28, 1983. Healed decubitus ulcer shown on foot on bottom photo taken on February 16, 1983 after thirteen one hour tonations.

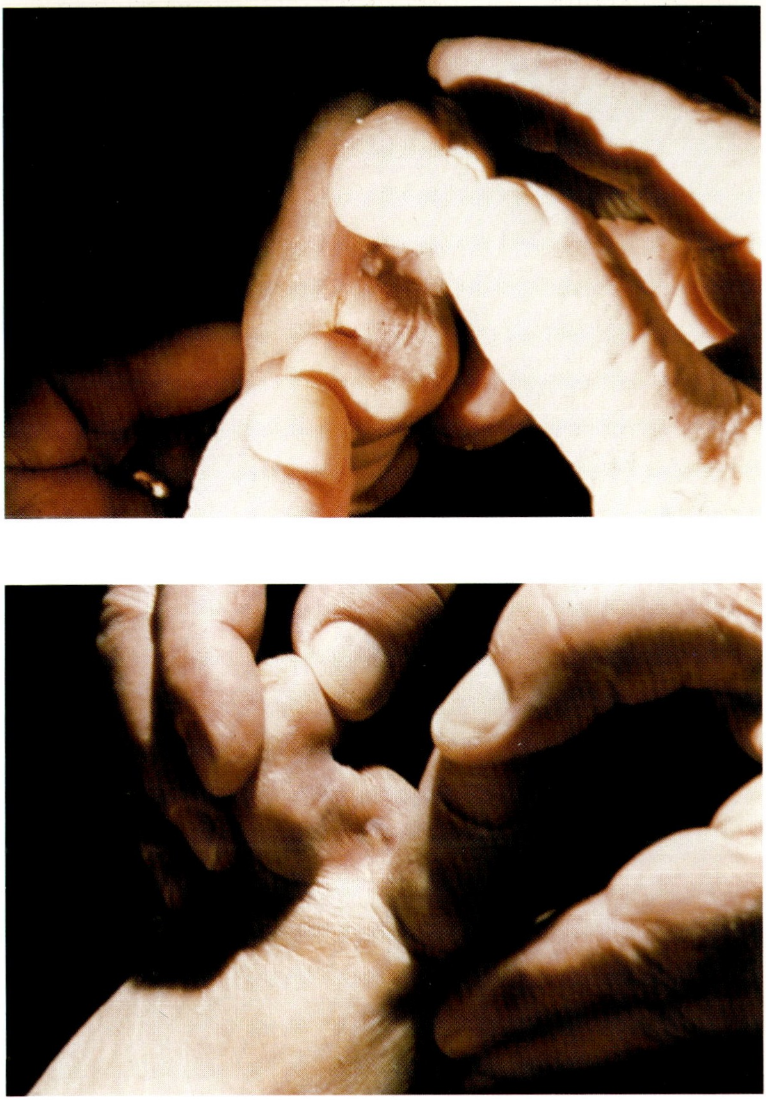

FIGURE 13

Onslaught of decubitus ulcer between great toe and second toe shown in illustration in top photo. Top photo taken at 10:30 a.m. on January 18, 1983. The bottom photo taken at 7:30 p.m. on the same day after one light tonation. Both times were witnessed by a registered nurse.

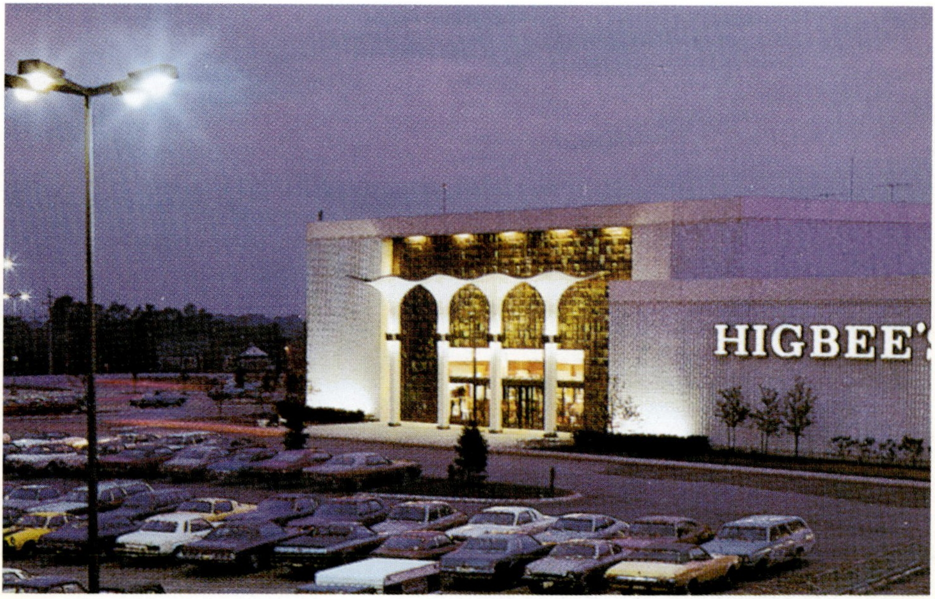

FIGURE 14

Low Pressure Sodium Light

PHOTO: GENERAL ELECTRIC COMPANY, NELA PARK, CLEVELAND, OH

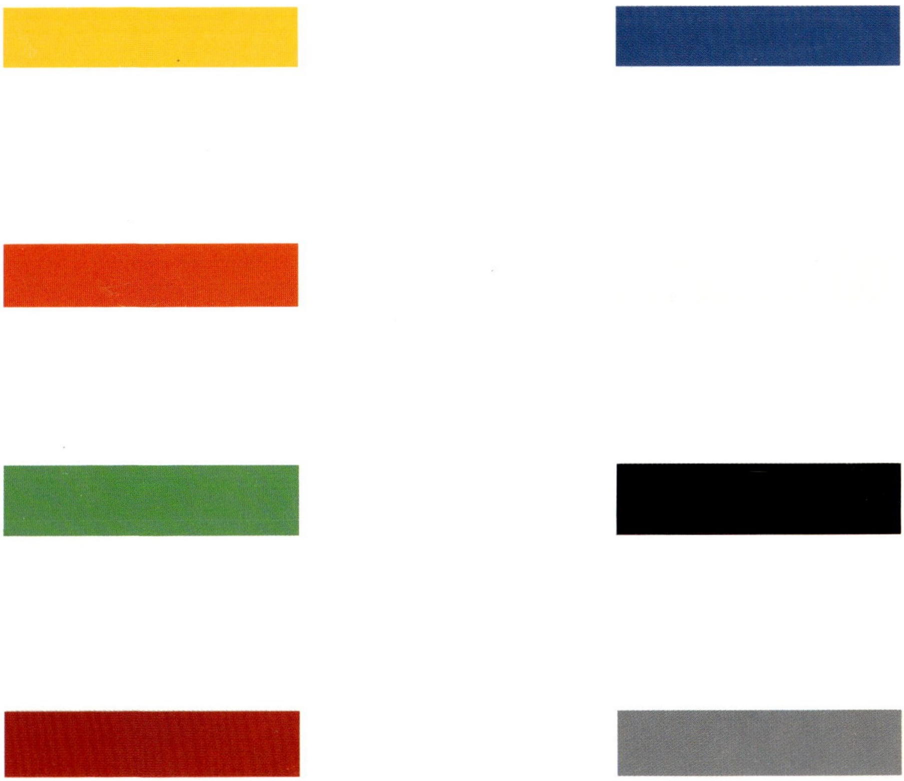

FIGURE 15

Safety Colors

PERSONAL COLOR KEY REFERENCE GUIDE

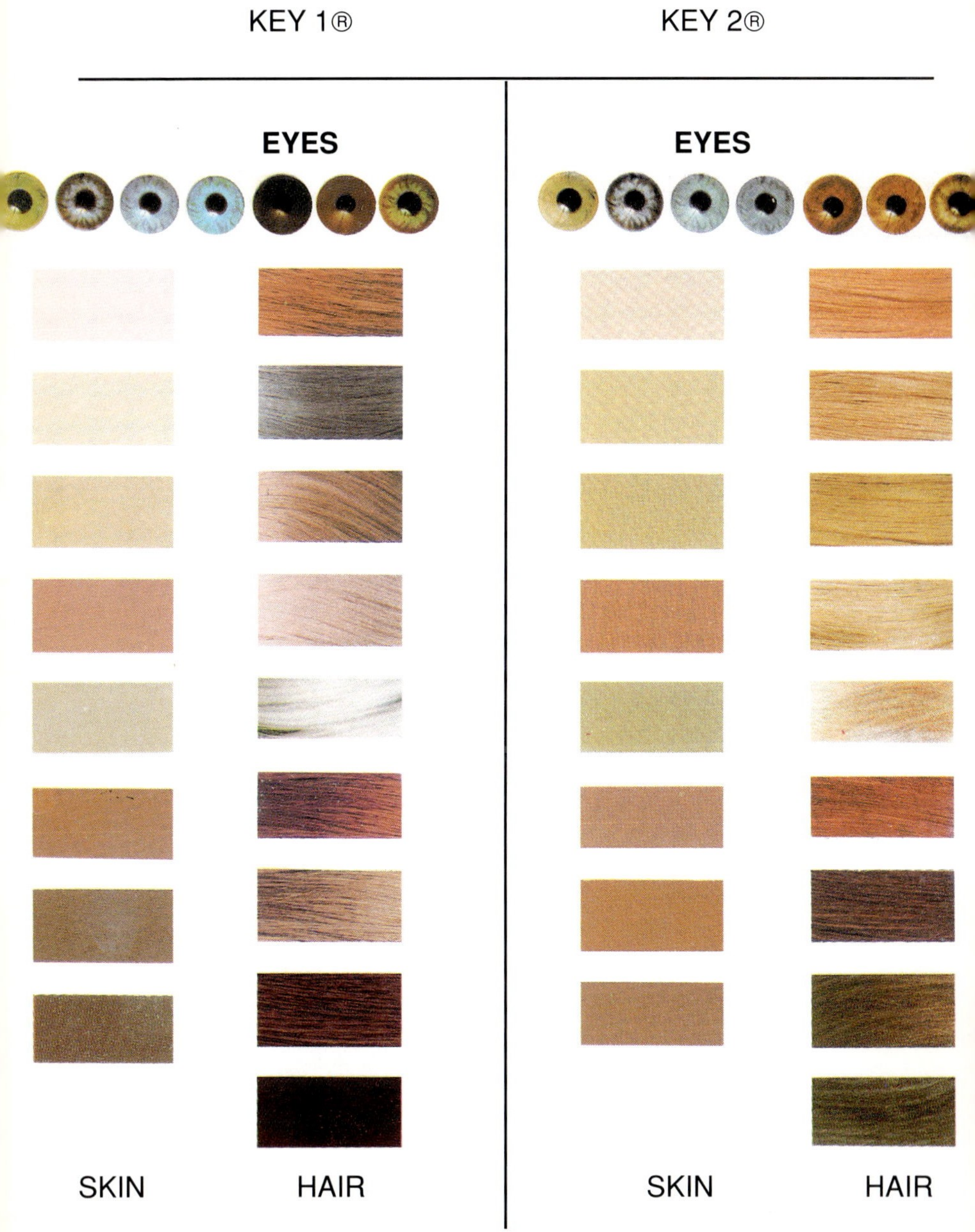

FIGURE 7

children and found they experienced clearer vision and were much less subject to fatigue. The use of full spectrum light is becoming more commonplace in medical settings, industrial and office environments, as well as our homes.

How people feel directly affects their life style, and therefore their quality of life. Performance in these areas can be enhanced through the correct use of light and color. The evidence is increasing —— Light and color relate to more than just illumination or aesthetic values in shaping our lives.

From the point of view of living, light and color are an investment in "the good life". From the perspective of working, being productive and comfortable, they deal with optimum conditions. For the retailer, the sovereign criteria at the bottom line is high volume and profit, which can be achieved through the correct use of light and color. Whether for home, office, or business, light and color influence our sleeping, wakefulness, emotions and health, thereby playing a major role in our every day living.

ORANGE

Orange is used for any acute hazard that is likely to crush, cut, burn or shock a worker. It is also painted around the edges of cutting machines and rollers. You will find this color on the interior of switch boxes and fuse boxes. It acts as a warning to those opening the boxes as it silently yells, "**Be Careful**!"

GREEN

Green has always been identified with safety and medical care. It is used to identify safety equipment and its location. First aid equipment, cabinets for stretchers, gas masks, medicine cabinets, all will be painted green with identifying letters in white.

RED

This color is associated with fire protection. It is reserved for marking fire protection devices and for the location of fire fighting equipment. You will find it painted in a band on columns or on a wall behind a fire extinguisher. You will also find it on valves or fittings for hose connections.

BLUE

This color says caution. A blue flag with white lettering is used to mark any type of equipment that is "out of order" or undergoing repairs. The railroad industry uses it to mark rail cars that should not be moved.

WHITE /BLACK/GRAY

White is used in square areas to mark where stock is stored. It is used in locations for the positioning of waste receptacles, and marking aisles. White baseboards and corners will discourage littering and encourage whomever does the housekeeping to clean along the edges and into the corners.

Having a consciousness about color safety increases awareness of hazards, which in the end results in lowered injury rates. Color some sort that will carry the electrical charges. The conducting fields, studied by scientists for years, are called “L-fields”. In 1932, a geo-physicist and engineer, Erich Konrad Muller, published his findings on electrical energy that emanated from the body in the English Journal Of Paraphysics. Then Dr. Harold Saxon Burr, a professor of anatomy and biology at Yale University, spent thirty years investigating electrical systems in all forms of life. He established through his studies fields that he called “L-fields”, or “fields of life”.

Faith healers say that they can feel these electrical fields. They experience a type of transaction between themselves and the person they are healing. Their experience is characterized by one of several effects that can be described as “tingling”, “pressure,” “pulsations,” and “hot” or “cold” sensations.

No matter how one intellectualizes therapeutic touch, or the human electrical field, it is a phenomenon brought about by our body’s own natural electrical action. This action creates a colored field that surrounds the exterior of the body due to the electromagnetic energy that is transferred. This field varies from individual to individual. When a person is calm, his/her field is small. Aroused to anger or passion, the electromagnetic energy field flares up and sparks with a brilliant red.

Musicians, healers, people in different walks of life and interpersonal relationships, all show a distinct difference in color and flare patterns that their bodies emit. Uri Geller, is well known for bending spoons and stopping clocks, without touching them, through his unusual energies. His fingertips were placed a few inches apart on a table top. A watch was placed on the table close by his fingertips, and as he concentrated on the watch, his fingertips were photographed with a Kirlian Camera. When the film was developed, you could see the evidence of shooting energy from his fingertips toward the watch.

This same energy seems to last forever. Radiations have been photographed on a mummy’s wrapped body that was thousands of years old. Additional evidence can be seen in a phenomenon in West Germany have proven that exposure to bright light at the right time of day is effective in shaking off the symptoms of jet lag.

Both Seasonal Affect Disorder (SAD) sufferers and those affected by jet lag are experiencing a disruption of their biological clock. The SAD patient's clock is running either too fast or too slow. Jet lag is created by geography and time differences confusing our systems. Our biological body clock lays down fundamental rhythms in our body to help it function. Included in these bodily functions is body temperature and the release of melatonin. Our bodily rhythms are synchronized by the earth's daily rotations around the sun. The one important thing that keeps the body's clock synchronized with the environment is a daily dose of daylight.

Several years back a study was done by scientists on the isolation of human beings in the dark. Volunteers spent weeks at a time in isolation at the bottom of a mine shaft. Their body cycle, based on standard twenty-four hour cycling, showed that they drifted away from that which is standard. After weeks underground, their body temperature rose and fell in a twenty three hour cycle, but more often it changed, based on a twenty five hour cycle.

That doesn't seem troublesome when you read about it, but conflict begins when your internal clock tells you one thing, but the outside world differs. This accounts for the jet lag experienced by those who fly great distances in a few hours. They disorient their internal clock. Our biological clock may call for behavior appropriate to early morning hours, while a cue from the outside world says it's high noon and sunny. The temporary disorientation is referred to as "jet lag."

The rods and cones of the eye are the photoreceptors of the visual system. The rods give sight by night. The cones give sight by day. Dr. John Ott, in his studies, indicates that the epithelial cells that lie directly behind the rods and the cones act as separate photoreceptor mechanisms. When certain wavelength frequencies are missing, the cells become sluggish and will not go through

CHAPTER 10

COLOR SYMBOLISM, PERCEPTION, AND PSYCHOLOGY

Because our perceptions of different colors are so strong, industry spends hundreds of thousands of dollars in test marketing products in packages that are color coded to gain your dollars in the market place. When a person makes a color choice, it is a subjective choice based on emotion, attitudes, values, and perceptions. Emotional values regarding different colors are bound to time, place, geographic location, culture, development, age, and gender.

COLOR SYMBOLISM

Many colors are linked to culture and historical times. In earlier times, the significance of color attached itself to great biological events, such as life, death, birth, and puberty. Painting the body in colored designs in primitive times, and even today among some tribes, indicated status, and told all who saw the colored designs that this individual belonged to a certain tribe or held a certain position within the group. New Guinea women from the Papua tribe cover themselves with clay that is white, to signify that they are in mourning. Talk to anyone of the western culture, and they will relate black as the color of mourning, or death. Go to China and you will see those in mourning dressed in white. Just from these few examples, you can begin to understand how strongly color is connected to culture.

effort to do so. You have the energy of red, and the spirituality of blue. Generous, sensitive, difficult to understand, and perceptive would be terms that accurately describe you.

Negative Characteristics

You are so interested in discovering and probing new ideas that what lies under your nose and eyes cannot be seen. As you dig deeper into the discovery of mysteries, your home piles higher with mess. Highly temperamental, prideful, and arrogant, these traits cause conflict with your positive side.

ORANGE

Positive Characteristics

Orange lacks inhibition, and sparks warm. You are a person who makes "small talk" with great ease. Sitting on a park bench, in an elevator, or with any stranger, small talk, an easy smile and a good social nature make you everyone's best friend. You'd make a great salesperson, teacher, or would be at home in any job where there is personal contact. Your energy is like that of red, but without the passion. You seek to please others and are agreeable; this helps to capture their attention and since you don't like being alone, helps to guarantee companionship.

Negative Characteristics

You tend to be overly conscientious and overly considerate. Because of your dependence on others, and because you strive to please and be considerate, you lose yourself in return.

FIGURE 5

FIGURE 6